RESEARCH ON TECHNOLOGICAL INNOVATION, MANAGEMENT AND POLICY

Volume 2 • 1985

RESEARCH ON TECHNOLOGICAL INNOVATION, MANAGEMENT AND POLICY

A Research Annual

Editor: RICHARD S. ROSENBLOOM
Graduate School of Business Administration
Harvard University

VOLUME 2 • 1985

 JAI PRESS INC.

Greenwich, Connecticut   *London, England*

CONTENTS

LIST OF CONTRIBUTORS

John M. Dutton   Graduate School of Business
                                Administration,
                                New York University

Karen J. Freeze   Graduate School of Business
                                Administration,
                                Harvard University

Jacob E. Goldman   Cauzin Research Associates,
                                Norwalk, Connecticut

Margaret B. W. Graham   School of Management,
                                Boston University

Cynthia Hardy   Faculty of Management,
                                McGill University

Andrew M. Pettigrew   University of Warwick,
                                Coventry, England

Richard S. Rosenbloom   Graduate School of Business
                                Administration,
                                Harvard University

Robert Stobaugh   Graduate School of Business
                                Administration,
                                Harvard University

Annie Thomas   Graduate School of Business
                                Administration,
                                New York University

INTRODUCTION TO VOLUME TWO

This book is the second in a series reporting current thought and research on the management of technology. The Introduction to Volume One stated our intention to address as audiences both scholars concerned with creating useful knowledge and practitioners in search of knowledge that can be put to use.

Volume One explored interrelationships of technology and competition. The papers in this Volume examine phenomena found largely within the firm. Their focus is on the institutional and organizational forces that shape the character of innovative activities. Volume Three, now in preparation, will examine the management of emerging technologies.

Several of the themes which link the several chapters of this Volume are introduced in the first, Jacob Goldman's insightful discussion of

"Innovation in Large Firms." Why, he asks, do large firms so often appear disadvantaged in innovation, despite the substantial resources they command? The basis of his conclusions, frankly acknowledged, is "anecdotal, rather than statistical." He brings to these judgements two decades of experience as Chief Scientist in major corporations, first Ford Motor and then Xerox Corporation.

The next three papers explore behavior in large firms. While Goldman calls attention to the "rigidities" of large organizations, Hardy and Pettigrew show how such tendencies can be overcome by the skillful use of power, sometimes quite unobtrusively, to smooth the way for change. Using the lenses of organizational theory, they examine phenomena associated with an inevitable concomitant of innovation, the obsolescence of the old.

An important institution in the world of corporate technology, the scientific research laboratory, is found only in the realm of large firms. Margaret Graham applies the skills of an historian to examine forces impinging upon these institutions in the postwar era. Each laboratory must find its own way to adapt to two major forces, the "climate" of science in the larger society and the "context" for research within the firm itself. Neither of these forces is static, and the way each organization works out its response to change will determine, in part, its ability to achieve its own mission.

Robert Stobaugh analyzes the nature of product innovation activities in an industry, petrochemicals, in which large firms figure as the dominant innovators. "How can large firms organize for innovation," he asks, "in order to take advantage of bigness while minimizing its disadvantages?" More than economic advantage is required, he finds, since the firm must be capable of managing a lengthy and risky interactive search process built on two complex communication networks— internal and external. His chapter outlines the way that this commonly occurs in an industry where innovation and profitability have long been closely linked.

Innovative small firms, if they succeed, must contend with a paradox of growth: to the extent that success leads to greater scale, it may negate the advantages of smallness that produced it in the first place. The chapter by Rosenbloom and Freeze recounts the history of a small high technology firm in California that succumbed to that paradox. Wildly successful as a pioneer in an emerging technology, Ampex Corporation became a much larger, more diverse organization, "pro-

fessionally'' managed, but ultimately unable to sustain its earlier innovative leadership.

Large firms, typically, have the advantage of ''experience.'' While
often cited as a source of competitive efficiency, experience is also a
source of knowledge that can lead to innovation. The final chapter, by
Dutton and Thomas, explores the state of our knowledge about the art
of ''learning by doing'' and its relationship to technological change.

Richard S. Rosenbloom
Series Editor

INNOVATION IN LARGE FIRMS

Jacob E. Goldman

Conventional wisdom attributes a significantly higher rate of innovation per technical employee to the smaller firm than to the larger firm. A study of innovation by the U.S. Department of Commerce in the nineteen sixties (the Charpie Report)[1] called attention to this disparity. More recent research on the subject not only confirmed this but suggested an ever widening gap between the large (more than 500 employees) and small firm and also a very significant acceleration of innovations to the marketplace by smaller firms.

To put it in more quantitative perspective, the Gellman study[2] examined 635 innovations that reached the marketplace after 1969. The study concludes that small firms produce 2.5 times as many innovations as large firms relative to the number of people employed and that small firms bring their innovations to market 27% more rapidly as determined by examining the mean time from the establishment of performance criteria to market introduction.

Research on Technological Innovation, Management and Policy
Volume 2, pages 1–10
ISBN: 0-89232-426-0

The proliferation of successful high tech companies in the last few years nurtured by the explosive venture capital market suggests that the disparity between the large and the small enterprise in innovation rate is, if anything, widening. The irony in this trend lies in that in the early post-war years it was generally assumed that the innovation proclivity—particularly in the U.S. economy—was fed by the industrial R&D establishments and these tended to be highly concentrated in larger firms.

The objective of this chapter is to identify those characteristics of the innovation process that tend to be inhibited in the large firm and conversely enhanced and otherwise supported in the small enterprise. The basis for judgment is anecdotal rather than statistical and is based on the author's experiences in managing the R&D function in several large corporations and by exploring through the venture capital and small business community the converse hypothesis.

It would be useful, first, to delineate the distinction between innovation and invention and, as a corollary, the innovation process as contrasted to the invention process. The familiar cartoon representation of the inventive spark as a bolt of lightning striking or an electric bulb flashing near the head of the inventor is not far off the mark. Invention can result from a flash of genius or painstaking pursuit of a technical response to an identified or perceived need—sometimes perceived only by the inventor himself. It is sometimes a product of gut feeling, intuition or internal correlation; in other instances it may result from the juxtaposition of carefully conceived scientific ideas. Examples of each abound. Most of Edison's inventions resulted from a remarkable intuition rather than deductive scientific reasoning. On the other hand, Carlson's invention of xerography, Bardeen *et al.*, transistor, Forrester's magnetic core memory and Land's instant photography belong more properly in the latter class.

Innovation, by contrast, is invariably the product of an extended series of steps that link the invention to the marketplace. Economics, manufacturability, reliability and reproducibility are important ingredients of the innovation process but are irrelevant in assessing inventiveness. Therefore, the very process of innovation requires sensitivity to these exogenous factors and will usually require significant time periods to gel while an invention can be instantaneous. Invention is usually the product of one or two or at most three fertile minds; innovation more often is the product of a team effort although on occasion the entrepreneurial instincts of one individual, the champion, whether or not he is also the inventor, can produce an innovation. A final distinction may also be appropriate: invention generally has a high technical component and focuses more frequently on product concepts; innovation, on the other hand, is just as likely to be completely non-technical (e.g., a marketing innovation) or, if technical, is as likely to be a process innovation.

A common misconception is that innovative enterprise is related to R&D within the enterprise. Perhaps this is so with regard to invention but not with regard to innovation. Patent statistics confirm a correlation between inventions and R&D; but one is hard put to find a relationship between R&D and the innovative capacity (or IQ—Innovation Quotient as Gellman calls it) of an organization. In fact, the thesis of this chapter is to suggest that notwithstanding the propensity to invention of establishments with competent and sometimes superior R&D organizations these do not necessarily produce innovations. We shall endeavor to understand why this is so.

The first area of significant difference between the large and small enterprise has to do with *organization* structure. In the large company, responsibilities are compartmentalized. It does not necessarily have to be so and there are some notable exceptions. (3M is one such exception and will be referred to later.) But by and large, creative innovations in the large organization are expected to issue forth from the technical side of the house—often the research and development laboratory. This immediately introduces a barrier that insulates the technical organization from the rest of the company. The barrier works both ways: it impedes the transfer of new technology to the operating groups, and impedes the flow of market information to the technical people. To be sure, many techniques have been discussed for overcoming this barrier—entire conferences and books have been devoted to this subject. (see, for example, J. Morton. *Organizing for Innovation.* McGraw-Hill, 1971.) But the existence of that barrier often accompanied by a geographical barrier adds a major resistance to the free flow of information between organizations charged with very specific responsibilities. The decision process involved in embracing new innovations in product lines or manufacturing processes involves many people whose goals diverge, thus complicating the decision-making process. In contrast, the small firm rarely exhibits the organizational rigidity that characterizes the large firm. The entrepreneurs and their founding colleagues invariably pool or share functions.

A corollary to the organization issue that distinguishes the large from the small firm is a *cultural* one. The organizational culture of the large firm inhibits the advancement of a new concept on toward the market-place—the conversion of inventions to innovations—until all the disparate functions are completely satisfied. There exists what one may call an organizational overhead in this process. Before all the organizational components will sign off on a potential product program, each will affix its own requirements, each will study the impact on price, market penetration, manufacturing cost, etc., of its added function. For example, product reliability and service requirements can be traded off by the engineering and service organizations and each will impose its own norms—not to mention added costs associated with maintaining the integrity of each of the components. There is, as a result, a loss of flexibility in the large organization and a compounding of overheads which translates itself into time delays in both the decision making process and the introduction to market. Often the process will rule out a product opportunity before it is even born. Quoting Utterback: "It's difficult to be flexible in an emerging technology in an organization that has been designed in a very hierarchical manner for high volume production, and where decisions center around allocating major capital and have to be taken with a good deal of care and consultation."[3]

Perhaps recitation of a specific example from the author's experience at Xerox will serve to illustrate this phenomenon. The Xerox Palo Alto Research Center was created in 1970 to provide Xerox with technological capability in the explosive world of electronic digital processing of information. Consistent with the corporate strategy of expanding into office automation, the computer scientists recruited into the laboratory turned their attention to marrying the traditional company capability in graphic communication to the potential of processing information digitally. Two product opportunities emerged very rapidly and by late 1972 the technical embodiments of both were completely in hand. One was a laser printer enabling the conversion of digital signals into high resolution, full graphics xerographic output. The other which became known as the ALTO was a full function work station with a digital processor architecture configured to optimize office functions such as text editing, electronic publishing, full range graphics and with radically new, well-conceived human factors which replaced verbal "computerese" instructions with image ikons and a "mouse" instead of a keyboard-operated cursor or light pen. A special powerful high level language was also developed for the machine. The in-house reception to the technology was staggering and by 1978 twelve hundred such terminals were in use throughout the company and in several universities chosen for joint experiments (particularly on software).

In terms of the then prevalent office environment the ALTO system was, indeed, an innovative approach, especially if tied to a laser output printer.

The key question that had to be addressed by the company at that stage (i.e., after many thousands of machine-hours of successful in-house experience in the hands of professionals as well as clericals) was how and when to take the product to market. There began a series of complex decision making steps that are so characteristic of the large firm. Clearly, the cost (or sales price) of a research-designed machine would be too high for the mass market. Laser printers weren't available for the external marketplace. Electronic printing was the responsibility of a division other than the office products division—which would have in the normal course of events the responsibility for bringing an office terminal to market—and a decision was made to use a new generation xerographic engine for the small and medium sized laser printer. Hence that xerographic machine wasn't even ready.

In the small entrepreneurial firm, the decision would have been easy: take the machine to market as it is, capture the limited price-insensitive market for which the feature richness of the machine would override the price barrier. The engineers would help support the hardware and software in the field, and feedback from the users would be useful in preparing for a second model which would also be cost reduced to command a broader market expansion. This was, in fact, the approach proposed by the research team responsible for the development and might have been embraced in companies such as 3M which tend to create small entrepreneurial enterprises built around the program initiators to promote new products.

But in the stereotyped large company environment that characterizes the Fords, Xeroxes, GMs of the world the path to market was strewn with far more obstacles. First the marketing organization (the Office Products Division) had to be persuaded that the product merited attention; they had their own product scenario based on by now obsolete technologies and there was the not unexpected reluctance to supplant a product stream that already had its champions. Then after extensive persuasion to accept the concept, they set specs and cost targets that required that the developers go back to the drawing board.

The cost targets were based on the firm requirement that the marketplace absorb a minimum number of placements significantly larger than the product concept as originally configured could command, and which could support the necessary training and upkeep of the sales and service force. Then the Electronics Division, which represented engineering and manufacturing for electronic components, got into the act and strongly urged a redesign of the processor configured in LSI with more limited power but presumably more economical. After extensive back and forth arguments, trade-off negotiations and management diversion, the product decision was finally made. But the cost and delivery date targets that helped persuade management to favor redesign were never met. The Xerox 8010, the ultimate product derived from the ALTO, was finally introduced to the market in the fall of 1981—a full five years after the more limited volume product was proposed for the marketplace and still far too expensive for the broader marketplace anticipated. During those five years competitive technology moved forward so that some of the innovative uniqueness was lost. In fact, in late 1982 Apple introduced the LISA which embodied a great many of the ALTO innovations at a much lower price. To be sure, the 8010 is a unique and superior product and may yet find its proper market niche and become a profitable product. But the large company syndrome stifled what should have been one of the most remarkable innovations in office automation of the 70s.

The ALTO example illustrates another disparity between the large and small organization which can be described in terms of *time horizons*. This is intended not so much to mean the time delays inherent in the complexity of the large organization but rather the perception of time in a multifaceted organization as distinguished from that in the small company. U.S. companies boast of their dedication to the concept of long range planning. Most companies have on their books five year plans and some even ten year plans. The problem is: suppose an innovation comes along either through internal development or from the outside that does not fit into the scenario as projected by the planners. The resistance to modification of the plan is not simply a manifestation of the NIH syndrome, but a natural reluctance to accept the domino effect of all other displacements necessitated by perturbations to the plan. This is not simply reluctance to accept new products; it manifests itself as well in refusal to acknowledge exogenous variables that could impact the orderly evolution of the "plan." Thus, the automobile industry of the sixties and early seventies refused to entertain the very thought that pollution, safety or prospective oil shortages should be factored into and produce modifications of plans writ in concrete by the financial and marketing gurus who determine the course of the industry.

The author in 1956[4] wrote a "what-if" scenario based on a projected oil famine 25 years thence. While the effects of that prognostication made their way into the research programs of the host company during the following decade, the makers and planners went their merry way with nary a thought to alternative planning assumptions. This is not a conducive environment to innovation. The dire consequences to the industry when these remote eventualities emerged are well known and the industry was left with neither alternative scenarios nor product entries. The small company does not have that problem because it can turn on a dime when conditions demand, but more importantly it can keep alive options that are likely to be foreclosed in the more rigidly planned environment of the large company.

The propensity of the small company toward innovation compared to the large is in a way quite surprising. As noted earlier, sophistication in research and development is the hallmark of the large company. Particularly in the case of basic research—the wellspring of the innovation process—it is virtually non-existent in the small company. Is it possible that the tendency to innovation does not correlate with R&D capability and, if so, why not? The record suggests that indeed there is little correlation between innovation (as distinct from invention) and R&D. The reasons are many and likely derive from the barriers mentioned above. An additional factor may lie in the circumstance that results of R&D are unpredictable. The innovative results of R&D may be applicable outside the bounds of the strategic product stream of the company. Unless the company is prepared to diversify into a new product opportunity, or, spin off a separate organization to develop and promote the product (or process), the new idea will die on the vine and the innovation will not emerge.

In recent years, the emergence of a powerful venture capital force has caused frustrated innovators to leave the secure shelter of the big company and create an environment more conducive to the promotion of innovations in the small start-up company backed by new sources of capital. Thus R&D does and will influence the innovative process, but the large company does not profit from it.

If companies do not wish to cook the R&D goose that lays the innovative golden eggs they will expand their corporate strategies to

allow for off-line exploitation of these eggs and eliminate the frustration of the R&D people who see the fruits of their efforts come to
naught. To be sure there are exceptions. Most notable is Bell Laboratories that has had a continuous record of innovations spanning nearly
three quarters of a century not only in the telecommunications industry
but extending well beyond it. They have been unique in translating
invention to innovation and marshalling the necessary inputs—
whether outside or inside the AT&T organization—to help promote
the innovation. Why they have been so consistently successful at it is
what Jack Morton's book—mentioned earlier—is all about.

Perhaps the example of the transistor will illustrate one particular
characteristic of Bell that catalyzes the innovation process. The story
of the invention of the transistor by Bardeen, Brattain and Shockley in
1947 has been well documented and chronicled. But apart from its
significance as a product of a well conceived R&D strategy, the steps
that led from this one *invention* to the revolutionary innovations of the
microelectronics industry which later coupled back and were duly
exploited by Bell are particularly noteworthy. By recognizing that the
advantage to Bell would be enhanced if others would play a part in
developing the technology and thus opening it up to broad licensing the
Bell management showed great foresight and perspicacity, all of which
was proven by subsequent events.

In a sense, the Xerox experience of broadly licensing Ethernet—on
the assumption that others may adopt it as a networking strategy which
would then benefit Xerox—was similarly conceived. Whether the results will be as far-reaching is questionable but the principle is similar.

These two examples are cited in order to emphasize that organizations with strong R&D capability can, under the right circumstances,
catalyze the evolution of inventions to innovations without compro-
mising either the integrity or the loyalty of their R&D organizations.
Clearly, these examples plus others in the pharmaceutical and electronics industries suggest that under the right set of circumstances the
large organization can succeed. The process of recognizing innovation
and nurturing the entrepreneur within the large organization is a difficult but not impossible one.

What organizations need is a willingness to experiment with alternative schemes instead of the rigid hierarchical structure of most organizations. At the General Instrument Company (on whose Board of
Directors both this writer and the Editor of this volume serve) the CEO
has created an imaginative mechanism to stimulate entrepreneurship
and reward the entrepreneurs by setting up independent companies

within the company complete with phantom stock, independence from corporate staffs and devoid of all the pressures and trimmings of the parent organization. Other experiments of a similar nature are known to the author and will undoubtedly proliferate in the coming years as more organizations recognize that they will lose market position and market leadership to foreign competitors or domestic entrepreneurs if they do not master the innovative process by a willingness to experiment, to nurture the innovators and loosen the organization constraints on the process.

Finally, a word about corporate management and its role in the innovative process. Corporations, like individuals, have personalities, cultures and traditions that influence the corporate decision making process. Corporate personality is most frequently derived from the attitudes, backgrounds and personalities of the Chief Executive and the group of senior executives who surround him. When the leaders have entrepreneurial instincts, the corporate risk-taking instincts will reflect it. Conversely, when the top management is rigid, conservative, driven by the numbers and the ROIs, one is not likely to see the element of risk taking that an innovative company demands. One doesn't see too many experienced entrepreneurs in the executive suite of the typical large corporation. More likely they are financial types—"bean counters" as they are pejoratively called—lawyers or salesmen. On balance that background is not likely to produce an entrepreneur or a leader who is prepared to bet on the instincts of his technical people, which is what entrepreneurship and innovation are all about. I know of one company where management was concerned about costs and schedules of its engineering organization and so named an accountant to the function of Chief Engineer. It wasn't long before communication broke down between him and some of the most creative and innovative engineers that had to report to him. Before long some were forced to depart—all in the name of organizational formality and discipline. That is a symptom of a kind of intellectual malaise that grips the latent innovative capability of a company when its leadership does not understand innovation or grasp the environmental parameters that catalyze innovation, leading to frustration of the innovators and their ultimate sterility.

NOTES

1. Technological Innovation: Its Environment and Management, U.S. Department of Commerce Report 0-242-736 (1967).

2.   Gellman Research Associates, Inc., Indicators of International Trends in Technological Innovation, 1976.

3.   James Utterback, Proceedings of M.I.T. Symposium on the Innovation Process, M.I.T. Industrial Liason Office, November 1981.

4.   J. E. Goldman, *Physics Today*, 10:25 (1957): quoted in *Business Week*, February 16, 1976.

THE USE OF POWER IN MANAGERIAL STRATEGIES FOR CHANGE

Cynthia Hardy and Andrew M. Pettigrew

INTRODUCTION

One of the biases in the way the literature on strategic decision making and change has developed is the clear preoccupation with the analysis of the content of strategy to the virtual exclusion of theorizing about the processes by which strategic changes are formulated and implemented. Worse still authors such as Andrews (1971), Hofer and Schendel (1978), and Porter (1980) express their ideas about the content of strategy with an implicit process theory which assumes a rational theory of choice and change long since discarded by process analysts as being divorced from the empirical reality of how decisions

Research on Technological Innovation, Management and Policy
Volume 2, pages 11–45
Copyright © 1985 by JAI Press Inc.
All rights of reproduction in any form reserved.
ISBN: 0-89232-426-0

and changes are actually made. Thus as applied to the formulation of strategic change the rational approach describes and prescribes techniques for identifying current strategy, analysing environments, resources and gaps, revealing and assessing strategic alternatives, and choosing and implementing carefully analysed and well thought through outcomes. Depending on the author, explicitly or implicitly, the firm speaks with a unitary voice, can be composed of omnipotent, even heroic general managers or chief executives, looking at known and consistent preferences and assessing them with voluminous and presumably apposite information, which can be organised into clear input—output relationships.

However, during the 1970's authors such as Bower (1970), Allison (1971), Pettigrew (1973), Mumford and Pettigrew (1975), March and Olsen (1976), Mintzberg (1978), and Quinn (1980) have started to redirect research on decision making and change from its prevailing concern with rational analytical schemes of intentional process and outcome, and see decision making and change in a variety of process modes. These process theories are now familiarly expressed in the language of bounded rationality, incrementalism, politics and power, and March and Olsen's (1976) garbage-can frame of reference for thinking about organisational problem solving. Thus Bower (1970) conceived of strategy developing at multiple levels in the organization through processes characterized as chains of commitment leading to eventual confirmation. Phase metaphors describing the life history of investment proposals from definition to impetus were postulated and aspects of the impetus phase were described in political terms. Top managers were seen to determine the structural context through which capital proposals would pass and be filtered. Mintzberg (1978) conceived of strategy as consistency over time in a stream of decisions. This approach allowed him to distinguish between intended and realised strategy, and to pinpoint strategies contributed after the fact, or as he puts it realised despite intensions.

Another of Mintzberg's contributions was to postulate both that strategies appear to have life cycles—period of incubation, development, and decay, and that distinct periods of change and continuity can be discerned in the overall pattern of strategic development of the firm. This notion that firms may have patterned and periodic shifts in their strategy, Mintzberg and Waters (1982) has been extended by comparative case study research conducted by Miller and Friesen (1980) using the metaphors of evolution and revolution, and by recent historical work on firms in the pre-nationalised British steel industry (Boswell 1983), and by a comparative and longitudinal study of strategic change in the four largest divisions and corporate headquarters of Imperial Chemical Industries (Pettigrew 1985). Boswell (1983) identifies what he calls phases of strategic concentration around values of for a time growth, and then efficiency; and Pettigrew (1985) indicates how and why strategic changes tend to occur in packages interspersed with long periods of incremental adjustment. Such periods of high levels of change activity when strategic reorientations occur, are often triggered by major environmental disturbances, but the actual character of the changes which occur in the crisis periods are significantly influenced by long run attempts to change the dominanting idea systems and power balance in the firm in the periods leading up to the 'window for change' enabled by the environmental crisis (Pettigrew 1985).

In summary, this empirical process research on strategy has made a number of descriptive contributions to the understanding of strategic decision making and change. Strategic processes are now accepted as multi level activities and not just the province of a few, or even a single general manager. Outcomes of decisions are not just a product of rational or boundedly rational debates, but are also shaped by the political interests of and commitments of individuals and groups, the forces of bureaucratic momentum, gross changes in the environment, and the manipulation of the structural context around decisions. With the view that strategy development is a continuous process, strategies would now be thought of as reconstructions after the fact, rather than just rationally intended plans. The linear view of process explicit in strategy formulation to strategy implementation has been questioned, and Chandler's (1966) dictum that structure follows strategy has been modified by evidence indicating why and how strategy follows structure (Galbraith & Nathanson, 1978; Pettigrew, 1985).

However, useful as this process research has been in descriptively analysing how significant changes are made, with the exception of Quinn (1980, 1982), Kanter (1983), and Pettigrew (1985) few of these process researchers have tried to build on their descriptive analyses and pinpoint some of the key prescriptive features of managerial strategy required to create substantial changes in manpower, structure, or strategy. Quinn's (1980, 1982) discussion of logical incrementalism is one plausible managerial view of strategic change. Quinn describes and prescribes a process of strategic change which is jointly analytical and political and where executives are recommended to proceed flexibly and experimentally from broad concepts to specific commitments. A cautious step by step activity of building awareness of the need for change, legitimizing new viewpoints and challenging old assumptions, making tactical shifts and finding partial solutions, overcoming and neutralizing opposition whilst building political support around particular ideas, and then formalizing commitment for action. Muddling through with a purpose as he puts it.

Kanter (1983) on the other hand sees a prototypical innovation having three waves of activity occuring in sequence or as successful iterations. First, information is acquired, sorted, and exchanged to shape the definition of a problem which may become the focus for change. Then coalitions are built, teams created, and individuals encouraged to buy in or sign on for the change in question. Finally there is a mobilization and completion phase in which the boundaries and momentum for the change are maintained, opposition and interference is dealt with and the change proceeds through periods of secondary and subsequent redesigns until particular pieces of the change are implementable. Kanter (1983) sees champions for change ideas holding together and managing the above three stage process. Using what she calls power skills to persuade others to invest support and legitimacy in new initiatives; team skills to share information, resolve differences, and generate enthusiasm and commitment to particular solutions; and change architect skills to design and construct micro-change which are eventually connectable to macro-changes or strategic orientations.

This chapter seeks to build on previous work by Pettigrew (1975) (1980) (1985), Quinn (1980) (1982), and Kanter (1983) emphasizing the role of power and political processes in managerial strategies for creating change. Using two case studies of management attempts to close down factories, the chapter describes and analyses how environmental and managerial explanations can be profitably combined to analyse the processes of successfully managing cutbacks. In exploring the managerial strategies for implementing such changes, strong emphasis is given to the use by managers of unobtrusive power sources in order to legitimate cutback decisions in the eyes of employees, and thus prevent resistance and opposition to the closure of factories.

Cutbacks, in a variety of forms, have become a relatively common phenomenon in many different kinds of organizations in the western world. In contrast to the growth and affluence that followed the second world war, contemporary economies are more likely to be characterized by falling employment, cuts in government expenditure, lower birthrates and smaller markets. These trends have affected all sectors of the economy. In the private sector, with which this paper is concerned, the result has been reductions in manning requirements, the rationalizing of operations, and the closing of factories.

The current prevalence of decline has not always been matched by a concomitant research interest (Whetten, 1980). There has, in the case of factory closures and redundancies (permanent layoffs), been a significant amount of research but the bulk of this has focused on how the redundant individual subsequently fares in the labor market: reflecting the labor economist's perspective (Wood & Cohen, 1977/8). In contrast there has been a relative paucity of studies examining how closure and redundancy decisions are implemented. We need, in effect, to look at how redundancy and closure are managed, particularly if we are to fully understand and explain the different ways in which workers have responded: ranging from concerted resistance to factory closure, on the one hand, to workers volunteering for redundancy even in areas of high unemployment (White, 1983).

This paper examines two factory closures which were marked by an absence of overt opposition by employees.[1] It describes the way in which management implemented these closures and then offers a number of explanations for the lack of resistance. One of these explanations—the use of power by management—is then examined in more detail.

In both the case-studies documented below interviews were carried out with a number of individuals, including managers, shop stewards, union officials, employees and ex-employees. "Key" actors (i.e. those identified by other actors or by minutes of meetings as playing an active role in the closures) were interviewed first, followed by a selection of passive actors who had been affected by the closure without playing an active role in it. With regard to the latter, an attempt was made to talk to representatives of the various interest groups, for example, white collar, blue collar members of the different unions, employees still working for the firm, employees who had left the firm. In total some 30 individuals were interviewed in each case-study. Interviews were in-depth and unstructured, between 1 and 3 hours. In addition documentary evidence was examined: reports and memoranda supplied by both management, union, and newspaper articles. These had the advantage of providing contemporary evidence that could be used to validate the retrospective data of the interviews, carried out in

1978–80. The following section provides a summary of these data in the form of a description of the events at the two closures.

THE CAMERONS CLOSURE

The first case-study concerns the closure of factory in the north west of England. The factory called Mountside Works is a production site in one of the divisions of a large British multi-national, known here as Camerons. Mountside Works is situated in a large city in the north-west, on a site adjoining divisional headquarters. The factory manufac-tured a large number of chemical products, most of which were inter-mediaries and sold to other sites in the division and outside the com-pany. There has been a long history of chemical production at Mount-side—chemicals have been manufactured there in one form or another for 100 years and the company has been involved for over 50 years.

In 1975 the divisional chairman announced that Mountside Works would close in the next five to ten years. This decision can be traced to a number of events. First, the last piece of major investment had been in 1952 and, although maintenance work had been undertaken, the works and its equipment were old. Second, in 1971, the divisions of Camerons had been reorganised and the nylon business, previously a big moneyspinner for the division, had been moved elsewhere. This immediately highlighted the difficulties being experienced in the re-mainder of the division's business and profits had begun to fall. With the inception of a new chairman in 1971, studies had been undertaken to improve profitability by eradicating and rationalizing the products which were losing money. A third problem was that European com-petitors were working from a small number of highly complex inte-grated sites, whereas this division had ten production sites and, conse-quently, the rationalization of these became a priority. Within the terms of this policy, Mountside was not a good candidate for survival: it was wedged in between divisional headquarters and a hospital with little room for expansion; it was situated in a built-up area; and with its aging plant, the costs of modernization would have been high. These problems are all the more apparent when it is realized that there were already large integrated sites elsewhere in the division which were, in the eyes of one divisional manager, "the obvious places to concentrate production."

Following the 1975 statement, the assistant works manager embarked on the preparation of a closure plan detailing how the numerous products and businesses should be run down. Alongside this business plan was a "people plan" stating how many people would be needed, of what skills, and for how long. The rundown was to take five years to marry both business and people needs. The former required time to allow the transfer of some products to other sites where new facilities had to be built; in the meantime Mountside was to keep on producing. Employee needs also dictated a gradual rundown in order to disperse the release or redeployment (transfers to other posts within the company) of more than 1000 individuals.

A year later, in March 1976, a second announcement was made, stating that Mountside Works would close by the end of 1980 with the rundown commencing immediately. The works manager—John Oppen—made a presentation of the reasons behind the closure, and details of the rundown to managers, supervisors, shop stewards and elected representatives who, in turn, passed the information on to the rest of the workforce. Union officials were informed by letter.

The shop stewards immediately began to think of fighting the closure and contacted their local officials and the Member of Parliament (MP). Oppen then made a presentation to the union officials in June 1976, explaining the closure and the provisions that would be made for employees in terms of redundancy compensation and redeployment opportunities.[2] After the meeting, the local officials passed the matter up to their national officials to see if anything could be done to prevent the closure. This is the normal procedure in Camerons—if agreement can not be reached between unions and management, the matter is passed up the hierarchy. In November 1976, the local officials had another meeting with Oppen in which they said that, after having discussed the matter with their executives, they were satisfied with the company's explanation and that nothing could be done to prevent it. They then said they intended to play no part in the local negotiations which would be left to the shop stewards to "get the best deal they could for their members". The stewards could, of course, call them back in if there was any failure to agree on a particular issue. This effectively marked the end of any talk of resistance on the part of the stewards.

In December 1975 a redeployment manager had been appointed to deal with the problems of those individuals who wished to be re-employed within the company. By mid-1976 a redeployment steering group (RSG) had been set up to discuss such issues. It consisted of the redeployment manager, a personnel manager, an assistant personnel manager, three craft stewards and three general (unskilled) worker stewards. This was a problem-solving group, not a negotiating body, whose aim was to identify problems and make recommendations to the works manager who would then make the final decision. The RSG had been established to include the stewards in the rundown and to deal with the particular problems of redeployment. In 1975 questionnaires had been sent out and had indicated that 83% of employees wanted to stay with the company, while only 17% wanted to take severance. As it turned out, however, only 37% of manual workers and 44% of staff workers actually opted for redeployment, the rest took severance. The problem of the RSG consequently became one of how to effect redundancy rather than how to achieve redeployment: finding arrangements whereby people could leave earlier than their scheduled departure without forfeiting their redundancy payments.

The people who were interviewed (between 1979 and 1980) expressed the feeling that, although the closure had obviously created some uncertainty, they had been satisfied that they would not be "thrown out on to the streets." Virtually all were happy with the redundancy terms, especially the older individuals who would also qualify for their pension. Many harboured the idea of getting a part-time job which, with their pension, would bring nearly as much money as they had been earning. The company also provided some help for those people who were trying to find new jobs: retraining schemes such as heavy goods vehicle driving, were provided free of charge; a "job shop" was set up with vacancies posted, and news of successful redeployments; time-off without loss of pay was allowed to attend job interviews, and advice was given on how to go about a job search.

The eventual result was, by management's standards, a success: the rundown schedule was adhered to and the closure was carried out as planned, there was no industrial action; and morale and productivity

Table 1.

	Output			
	1975	1976	1977	1978
Total Output tonnes	2600	2850	2500	1300
Output/40 hour man week—tonnes	3.55	3.5	3.33	4.26

Table 2.

Absence Statistics		
1976	*1977*	*1978*
8.17%	8.05%	6.24%

Source: Figures supplied by the Company.

were relatively unimpaired, as Tables 1 and 2 show. Furthermore the interviews uncovered no bitterness among the ex-employees—although regretful of the closure, they continued to respect the company.

It was a shame it had to happen but if it had to happen I don't really think that it could have been done much better. If it had to happen it was a success (ex-staff employee).

I could never call the company: I was pleased with the way it was done (ex-manual worker).

THE ANDERSONS CLOSURE

The second case-study concerns the closure of a factory in the west of Scotland: the factory—Newlands—was part of a non-British multinational, known here as Andersons. The factory was first opened in 1949, employed around 1500 people, and covered the general area of mechanical engineering.

In September 1978 it was announced that following the company's world-wide loss of $145.5 million for the nine months ending July 31st 1978, a study was to be conducted into the European manufacture of a particular product (product A). The terms of reference of the study were: given that the company had excess manufacturing capacity at its two European plants at Newlands and St Bertin in France, what would be the most efficient manufacturing arrangement for the company and what would be the best utilization of any manufacturing capacity that was released. The study was, in effect, to assess the feasibility of reducing the existing dual sourcing arrangements to a single source, in a bid to save money.

The recommendations of the study were presented in November 1978. They proposed the complete transfer of product A to St Bertin to save $10 million on overheads. A second product (product B) could be transferred from St Bertin to Newlands to save 500 of the 1500 jobs at a cost of $2 million. The latter recommendation was, according to the company, in recognition of its social obligation to its workforce and the local community.

These recommendations were presented to employees and union officials. These groups responded quickly to the news of possible job losses: shop stewards formed a joint union coordinating committee (JUCC) consisting of nine representatives of both manual and staff unions. They commissioned, through the Scottish Trades Union Council (STUC), studies of the commercial basis of Anderson's decision and of the social implications of the layoffs. They lobbied politicians and other unions, they organized a demonstration in London, and a visit to the French factory.

In January 1979, the stewards and unions presented their case to the company: the report they had commissioned from professors at a local university argued that product A should continue to be partly sourced at Newlands, and product B should remain in France. They also pointed out, on the basis of the other study, the severe effect the redundancies would have on the local economy. The unions, however, had not been allowed access to the company's figures on the grounds of confidentiality; instead they had been forced to make various assumptions. These assumptions were contested by Andersons who, at the same time, were unable to "prove" their own assumptions without releasing these key figures. The result was that the series of discussions eventually petered out, effectively having achieved nothing.

In July 1979 the company announced that the study was to be reopened and the deadline for the final decision was extended. This was followed by another series of meetings between the president of the company, the STUC, unions, stewards, and the local council, and also between the U.K. managing director, local MPs' the under-Secretary of State for Scottish Industry, and the Secretary of State for Industry. These meetings revolved around the possibility of saving jobs. In November 1979, however, it was announced that following the collapse of the market for product B, there would be a complete closure of Newlands, with the loss of 1500 jobs in February 1980.

In the meantime consultants who had been brought in to work out new payment systems for the possible arrival of product B, had been liaising with both managers and stewards, as a result of which an idea had emerged for bringing alternative work into the factory. An international work-search had been conducted to try and find a buyer for the factory, and a local search had been undertaken to see if there was any demand for the skills of the employees. The international search failed but local efforts established a small sub-contracting engineering firm. Management provided human and financial resources to help set up this new firm and a new company—Merryvale Ltd.—was born.

In February 1980 Andersons closed the Newlands factory without incurring any form of industrial action. It left Merryvale Ltd., which now employs around 200 people. Andersons is still largely responsible for the project but wishes to hold no more than a 20% share in the long term.

Both of these closures were characterized by the absence of industrial action by the workforce, and in the case of Camerons there was also almost a complete absence of bitterness. This might be considered surprising given the obvious negative effects of closures: employees lose jobs and, given the current economic problems, may find it difficult to find new ones. At the very least they will be forced to change their employment and cope with that upheaval. Union officials lose members, and the community loses a place of employment which can contribute to increasing unemployment in the area. As a result of these factors one might expect resistance to be more likely than acquiescence. If that is the case, the acceptance of these closure decisions and the absence of any overt confrontration over them warrants closer examination.

EXPLANATIONS OF ACCEPTANCE

One possible explanation might be that, despite any negative effects of closures and layoffs, *there is simply no history of worker resistance.* That is not the case, however, in the U.K.: the early seventies saw a concerted effort on the part of employees and unions to prevent the increasing number of factory closures, using a new form of resistance. Prior to the 1970s there had been no significant historical record of factory occupations. Between 1970 and 1975 there were some 200 "work-in" or "sit-ins" against proposed closures, involving 150,000 workers (Clarke, 1979). This form of protest appears to have originally been used in Britain in response to an attempt to close some Scottish shipyards in 1971 and this provoked similar actions in other factories (Hemmingway & Keyser, 1975). Since the mid-seventies there appears to have been a decline in the use of the occupation in the private sector, but it has become a powerful weapon in the health sector where a number of planned hospital closures have been delayed or prevented by nurses, doctors and ancillory workers (Vulliamy & Moore, 1979; Carpenter, forthcoming). In addition resistance has also occurred in the form of lobbies, demonstrations and threatened strikes by, for example, steel workers and coal miners.

There is then, in Britain, a significant tradition of resistance to closure and, for workers faced with redundancy, the existence of prominent role models.

A second explanation might be that *the unions and employees in these particular case-studies lacked the necessary power to resist the closure*. The sources of power necessary to contest closure are much the same as those necessary to contest any form of managerial activity. One obvious source of strength lies in a highly unionized workforce. Hyman & Fryer (1975) argue:

> Where a significant amount of the relevant employee group are outside the union and unlikely to follow its policies and instructions, or where there exists an alternative force of non-unionists the ability to exert effective pressure on the employer is considerably reduced (p. 162).

Unionisation is only likely to prove a source of power for employees if they have official support for their actions. The difficulties in contesting managerial decisions without such support has been noted by a number of researchers (see Matthews, 1972; Beynon, 1973; Beynon & Wainwright, 1979; Red Notes, 1978; Ursell, 1979; Bryer *et al.*, 1982). Support on a wider basis can also provide power, for example, from national union bodies, the Labour Movement, the media, and the public.

Another determinant of power, according to Hyman & Fryer (1975), concerns the strategic position of the workers in question—how easily they are able to disrupt production and exert pressure on the employer. Thus, strikes by print workers are often highly effective because of the perishability of their product, or by assembly line workers because they can bring the whole factory to a standstill. Pizzorno (1978) has argued that workers have no such power in the case of closure because there is no longer any demand for their product and, consequently, a strike is an ineffective weapon. While this may be true of closures of firms which consists of only one manufacturing unit (Lane, 1981), it is not the case where an organization is only closing one of many units.

Here demand for the product still exists and industrial action can be used to disrupt its production—so a multi-plant structure may be a source of power.

The traditional weapons used to fight closure have been strikes and occupations. Both of these will be more effective if the employees concerned are unionized and supported by their union and other groups. Strategic leverage will be increased if there are surviving plants. A strike will disrupt production, particularly if it extends to other factories, and an occupation can be used to prevent the disposal of assets (Greenwood, 1977) or inhibit the transfer of production.

A lack of these power sources may, to a certain extent, characterize the situation at Camerons: although blue collar workers were unionized, official union support was not forthcoming. This was partially due to the fact that the more militant union official—the local secretary of the Amalgamated Union of Engineering Workers (AUEW)—was ill and played no part in the closure, leaving the more moderate secretary of the National Union of General and Municipal Workers (NUGMW) as the leading local official. Another reason stems from the fact that, at the time the closure was announced, the local unemployment rate was only about 5–6% (compared to over 12% today): most officials were of the opinion that their members would have little difficulty in finding new jobs.

> You've got to remember that the climate is totally different today (in 1980). We had no unemployment then, or very little. Certainly as craft unions we had no major problem (local union officer).

Finally, local union officials were more concerned with the firms that gave the minimum statutory notice and redundancy compensation.[3] They felt in contrast, that Camerons was behaving in a fair and generous manner.

As a result of the local officials' lack of support for resistance the stewards felt relatively powerless:

> We were told then that there was nothing more we could do except get the best possible terms for the employees (steward).

> If the union officials were like that—what chance did we stand (steward).

The stewards also felt that there was little chance of other factories participating in a fight against the closure.

> The other factories wouldn't have come out for us (steward).

They were even dubious about the prospects of their own members becoming involved in industrial action.

> You can't put up a fight when the members won't be with you. A lot of members would say: 'What the hell's it got to do with you—you're stopping me from getting my redundancy pay' (shop steward).

The workforce had rarely engaged in industrial action prior to the closure: employees could remember, at most, only one strike in the previous twenty years.

> People don't want to go outside the gates—they don't even want to lose over-time (steward).

And even if the workers could be motivated to take action, the stewards were not convinced that it would do any good:

> they would have laughed at us if we'd gone outside the gates—they wanted to close it and you're doing it for them (steward).

On the other hand, the stewards did possess enough power to represent a threat in the eyes of management. This stemmed from the fact that the manual workers were unionized, and members of large national unions which covered all factories in the company. Furthermore there was a shop stewards committee in existence, which brought together stewards from different factories in the division: it would make an ideal mechanism with which to extend any action to other sites. Management were clearly worried about this:

> throughout the whole thing it was made very clear to me that my job was to contain it because if anything went wrong with the closure—whether it was a strike, or a big hold-up, or a disruption of any sort—it could very readily spread to the rest of the division with economic disaster (works manager).

At Andersons, the unions possessed considerable power that could have been used against the closure. The factory was almost completely unionized with regard to both manual and white collar workers. Most manual workers were members of the AUEW, while white collar workers either belonged to the Technical, Administrative and Supervisory Section of the AUEW (TASS) or the Association of Profes-

Table 3. Percentage of
Available Hours Lost
Due to Industrial
Action

1975	0.16
1976	3.88
1977	18.04
1978	0.45

Source: Figures supplied by
Company.

sional, Executive, Clerical and Computer Staff (APEX)—all of which were national and fairly powerful unions. Furthermore the stewards of the different unions were working closely together as is evidenced by the JUCC, and they had the active support and involvement of the officials at both national and local levels. Nor was industrial action uncommon at the factory as the above table shows, and there was the possibility of it spreading to other parts of the company as the stewards had obtained pledges of support from their counterparts in other British factories.

In both cases then, union/employee groups did possess some potential sources of power. Obviously it is impossible to tell whether these would have been sufficient to prevent the closures had they been mobilized. but what is more to the point is: why there was no attempt to utilize these power resources?

An explanation of this might be that *the continuance of the recession has served to dampen morale to the extent that workers could not summon the energy or the will to fight the closure.*

> It's difficult in the present economic climate to get support because everyone's looking over their shoulder about his own job. In industry generally the demoralization of the workers is quite dramatic (union official at Andersons).

> We've seen sit-ins, work-ins, pickets and demonstrations, but in the end firms have still closed down. We are trying a new tack. I suppose we could have got ourselves into a political argument, but where would that have got us (steward at Andersons).

While, this may, to a certain extent, characterize the situation at Andersons it is far from satisfactory and it does not really apply to Camerons which occurred earlier. Nor does it explain the fact that

Table 4. Age
Distribution of
Cameron's Employees
(%)

Under 18	0.6
18–21	3.2
21–31	8.9
31–46	39.6
46–60	42.3
Over 60	5.4

occupations were still occurring, for example in 1979 (at an Airfix-Meccano factory) and 1981 (by a primarily female workforce at a Scottish jeans factory) by workers with considerably less power than those at Andersons. Furthermore one might argue that as the recession deepens, jobs become harder to find, and workers have more to gain and less to lose by contesting closure decisions.

Another explanation why workers did not attempt to make use of their power is that *there was no need to resort to the use of power because the closure represented no threats and had no negative consequences.* One might argue that the provision of generous severance terms by both companies served to "buy off" the workers. It is undoubtedly true that many of the workers in both firms were long serving, relatively old employees (see Tables below) who stood to gain substantial sums of money.

Table 5. Service
Distribution of Cameron's
Employees (%)

Under six months	1.2
6–12 months	1.6
1–2 years	0.4
2–3 years	4.3
3–5 years	8.7
5–10 years	26.4
10–15 years	28.5
Over 15 years	28.9

Source: Figures supplied by the Company

Table 6. Age Distribution of
Andersons' Employees (%)

Years	Hourly Paid	Staff
16–24	5.08	3
25–34	16.58	9
35–44	22.66	17
45–54	29.09	35
55–65	26.59	26

It is also true the provision of generous redundancy terms helps to explain why the actual amount of people opting for voluntary severance at Camerons was much less than had been indicated in the earlier questionnaire.

> I am very conscious of the fact that the company's severance terms are so good that you really are bribing people to leave and they couldn't resist that bribery (manager).

> I reckon that with the money I get plus my pension I'll be able to take a part-time job—four hours a day—and I'll have as much income. So, instead of beating my brains out working eight hours a day, five days a week and sometimes more, I might live a lot longer and still have a reasonable standard of living (employee).

Employees at Camerons may have also stood a good chance of finding other jobs as the local unemployment rate was no higher than the national average (see Table 8).

Redundancy payments do not explain the situation at Andersons, however, where unemployment was considerably higher than the na-

Table 7. Service Distribution of
Andersons' Employees (%)

Years	Hourly Paid	Staff
less than one year	0.48	1
1–4	16.17	9
5–9	17.00	8
10–14	20.08	17
15–19	19.66	17
20–24	9.67	13
Over 25	16.84	35

Source: Figures supplied by the Company.

Table 8.   Unemployment Rates

	Local Rate (Camerons)	Local Rate (Andersons)	National Average
July 1975	3.4 (March)	5.9	4.7
July 1976	5.2 (March)	7.5	6.3
July 1977	6.9	9.4	6.9
July 1978	6.8	9.9	6.6
July 1979	5.9	9.3	6.1
July 1980	7.9	14.3	7.8

tional average (see Table 8). For these workers the prospects of finding alternative work were extremely dismal and their redundancy payments would not last long during a period of prolonged unemployment.

The provision of generous redundancy terms may explain why some workers accepted the closures, but it does not satisfactorily explain the quiescence of the workers at Andersons, many of whom might never work again. Nor does it explain the acceptance of the younger and short service employees who would make little financial gain. Another problem with this explanation, even at Camerons, is the commitment and attachment long-service employees often have to their jobs, and the dimmer prospects they have of finding alternative work. As workers stay with a company and get older they may have more to gain in terms of severance payments but, presumably, they also develop a deeper attachment to their jobs and their employers which could make changes all the more difficult, acting as a countervailing pressure.

MANAGERIAL STRATEGIES

The explanations documented above may partially explain the acceptance of the closures but, given their inadequacies which have already been pointed out, they do not appear to provide the full story. To do that we should examine more closely the strategies used by management to implement their closure decisions.

Managers in both the companies perceived a potential threat of resistance from employee/union groups and were anxious to avoid the negative effects in which it could result. They felt that workers had the necessary power to take industrial action against the proposed closures. At Camerons the existence of the unions and the shop stewards committee were worrisome:

> there is no doubt about it, at the time, we were very concerned about the amount of interchange between the stewards from one works to another (works manager).

> We were a little frightened of the union officials, particularly the AUEW officials (works manager).

At Andersons the existence of a militant, unionized workforce also gave cause for concern.

> We were never confident that it was going to go our way. We were never certain whether or not it was going to turn into an occupation: it was always on the cards (UK director).

Both companies were adamant, however, that resistance should not be allowed to occur. At Camerons the emphasis was on a smooth closure. It was

> In the minds of senior management that it (the closure should happen in a careful and relatively painless way (divisional manager).

This was because the five years rundown involved careful scheduling to enable the smooth transfer of the production of some products to new sites: industrial action would easily disrupt this.

> We had considered the possibility of an outright strike. That could have been quite damaging at that point because we had large chunks of the division's business, some of it perhaps not very profitable but nevertheless important, and important from the customer's point of view (works manager).

Furthermore if the action spread to other factories the division's business would be endangered. Camerons also wanted a peaceful rundown because they anticipated future closures and did not want to set a bad precedent. At Andersons the U.K. directors were also anxious to avoid resistance as it would have reflected badly on the ability of the U.K. management to handle a sensitive issue.

> It (trouble) would have hurt Andersons' image in the UK within the corporation. It would have shown that we couldn't manage a closure which America could do (U.K. director).

It was also conceivable that the entire U.K. operation could be threatened by industrial action. particularly if it spread to other factories.

> It was quite important for the future of Andersons in the U.K. that we could handle the problems without industrial action (U.K. Director).

Because of the perceived threat of industrial action and the desire to avoid it, both sets of managers felt it necessary to take steps to reduce the chances of confrontation. To do this they used a number of strategies to render the closure more acceptable in the eyes of their employees.

A.   The Creation of Managerial Credibility

In both cases managers tried to demonstrate their goodwill and create credibility for themselves and their actions. In Camerons this took the form of "climate-setting." In 1971 a new works manager had been appointed who immediately began to effect change at all levels in the culture of the works. This involved getting heads of departments to work together more closely, and bringing supervisors, managers, and later, stewards together. To do this residential courses were arranged for the managers and supervisors in 1974, bringing in the stewards in 1975. The works manager maintains that he would have introduced those changes in any event to improve working relations and to bring the works more in line with the rest of the division. Nevertheless, when Oppen was appointed he was aware that:

> In all probability my job was going to be to close those works,

and he did not believe that that would be possible to achieve, at least not in a smooth and painless way, with the authoritarian, cost conscious atmosphere that previously existed.

> I just can't see how we could have tackled the closure successfully in the old culture (works manager).

The culture change was fundamental in establishing trust and cooperation prior to the closure. It improved relations and increased consultation between managers and employees, particularly with the manual shop stewards who represented the biggest threat in terms of their ability and inclination to take industrial action. The result was that the stewards and employees developed far more faith in and respect for management.

> Management listens to you now: they didn't used to (shop steward).
>
> If you have a problem, there is always someone you can go to—someone who will listen (employee).
>
> I rather liked (the works manager), even though he did what he did (shop steward).

This type of environment was considered far more conducive to a smooth closure than the more autocratic climate which had preceded it.

> The previous works manager could not have done it because he was a different type of manager. I am not saying he was better or worse, but he was the wrong type of manager for this sort of situation (manager).

At Andersons the attempt to establish good-will has been categorized as tactical opportunism. U.K. management had been unable to create a favourable climate prior to the announcement of redundancies because they had learned about it from the company's headquarters only a few weeks before. They therefore had to rely on other methods to demonstrate their goodwill. This they did by taking advantage of a number of opportunities as they arose: in particular, the suggestion of an international search for a buyer, which was later expanded to include the local search for a market. As a result of the latter, Andersons then helped to set up a smaller enterprise on the site of the original factory. Management's goal was to:

> perpetually convince people that war (industrial action) would have a bad effect and they would lose out; that peace gave them hope and a chance (U.K. director).

These actions helped to accomplish this: they dissuaded workers from resisting the redundancies, and later the closure, in case it discouraged potential buyers.

> The one thing we couldn't afford was bad publicity because if you've trying to entice a company in and the company sees there's trouble, they're not going to come (union official).

It also gave management a chance to demonstrate their good-will and concern for their employees: the founding of Merryvale was presented as Andersons' duty to its workforce.

> But Andersons does not want to go into the sub-contracting business; it is merely shouldering its share of the responsibility for a community whose unemployment it has just pushed up from 9.8% to 12% (trade journal).
> [To a certain extent, this undoubtedly worked.]

> I believe that there was some sort of social conscience, if you can believe that in a multi-national company (shop steward).

> They accepted their social responsibility—they had a conscience (local council official).

B. Consultation

These attempts to create credibility for management were supplemented by extensive consultation with union representatives. At Camerons the RSG provided a mechanism whereby the stewards could be involved in the rundown process.

> Although 1976 was not a year which would present problems, the works manager saw the need around mid-year to respond to the wish of manual staff to be involved in the process of rundown, and . . . submitted a draft remit for a steering group (managerial report on the closure).

This provided a forum in which they could protect the interests of their members without having to directly confront management. The importance of this committee lay in its creation as a problem-solving group rather than a negotiating body. Issues could be discussed and solutions suggested, any recommendations were made to the works manager who could choose to implement or reject them.

> It can be concluded that the steering group was a massive time-wasting activity since many of the procedures it developed were little used, some convoluted arguments came to nothing in the end; nevertheless it was a very useful safety valve. It allowed the works manager to stand back from the fray and act as the final arbiter and so as the years rolled on what was achieved was a remarkable degree of trust on both sides. This prevented any real industrial unrest occurring or the need to resort to formal negotiating procedures (managerial report).

Andersons also embarked on an extensive process of consultation with employees, unions and stewards, the main reason for which was to avoid a procedural error:

to make sure that in no way could we be criticized for not having given the trade unions every opportunity to be consulted (U.K. director).

This consultation led to a series of meeting in which union officials met with management to discuss the closure. It was felt that there was a chance to prove to management on an economic basis that the closure was not necessary.

We felt that we could prove this was a viable plant (steward).

This resulted in energies being expended on proving that the case for closure was invalid, rather than fighting it through industrial action. Furthermore as the key figures were never released by the company, the unions were forced to rely on assumptions to present their case, the result being a rather futile attempt to disprove the company's arguments.

The focussing of our work on the accounts (underlying the closure decision) may have meant that other avenues were not explored. . . . We were using a method of argument which in the end we could never win. For every set of statistics we attacked, the company would come back and produce other ones. . . . It is possible that the contribution we made helped to reinforce, quite inadvertently, a strategy which, in the end, has produced the closure of that factory (academic who helped to compile the unions' answer to the closure decision, in a published article).

C. The Provision of Redundancy Payments

Redundancy compensation was also part of the strategy, having not only the advantage of "buying off" at least some of the workers: management, by offering more that the state minimum, was able, implicitly or explicitly, to threaten to withdraw it in the event of resistance.

We might have gone on strike but it wouldn't have done any good: we might have lost on redundancy (employee at Andersons).

If we had resisted they could quite easily have come up with the argument: all right, we'll give you the government's rate (steward at Camerons).

Redundancy payments are also important as legitimating symbols. Restraint programs quickly indicate to employees how highly they are valued by the company; redundancy compensation enables employers to demonstrate how fair and human they are (White, 1983; Luce, 1983).

D. Legitimizing the Decision

Both firms took steps to justify and legitimize their decision to close the factories. At Camerons the main reason behind the closure was the long-term plan to rationalize the number of manufacturing sites from ten to four, with production being concentrated on the larger integrated sites. The explanation of the reasons was not, however, always presented in these terms despite the works manager's attempt to tell the complete story. There had been a tendency for divisional management to emphasize the environment as having been responsible for the closure: the fact that the factory was situated close to a hospital and in the middle of a residential area, which could be potentially disasterous in the event of an explosion; the fact that the last piece of major investment had been in 1952.

> They (the divisional board) were tending to put around that we were closing the works because of the environment—it was a nice easy explanation (works manager).

Employees latched on to this explanation—out of eight stewards, ten manual works, and nine staff members, only two mentioned rationalization, the rest blamed the environment.

What is important about these reasons is not so much whether they were right or wrong, but the fact that they represented a convenient explanation which was accepted by the workforce as validating, justifying and legitimizing the closure, and that, to a certain extent, management was able to capitalize on this.

> Commonsense tells me it's not in a good position for a chemical works (staff employee).

> Everyone appreciated that it was an old works and changes were inevitable. They couldn't really spend the money to bring it up to date because it was so very old (manual worker).

The position of the factory even "delegitimized" resistance: it was perceived wrong to site a chemical factory so close to a hospital even though one had been there for 100 years, and the hospital had been built next to it.

> Everything was explained and it was found to be sensible, such as the hospital right on the perimeter of the works. When it was explained about the environmental problems, to even think about putting chemical products here wasn't commonsense (steward).

At Andersons management had argued that the closure was necessary to save overhead costs on a product that had been dual-sourced: by concentrating production in France which had ample capacity to meet current demand levels, the overhead costs of the Scottish factory would be saved. They were never able to justify the move to France but they were able to convince employees of the precarious financial position of the firm and the risk that industrial action might bring down the entire British operation.

> The unions were warned that if they pushed too hard there were 20,000 jobs in the U.K. at stake (manager).

> Nobody wants the company to collapse because there are a lot of people employed there (steward).

This also effectively delegitimized resistance.

The strategies described above appear to have helped managers in both companies implement the closure decision without incurring resistance. Naturally, it is difficult to say specifically how important they were in relation to the other explanations which have been offered: in all probability they all have some part to play and we may not be able to dismiss any one of them outright. But in response to the question: would industrial action have been more likely without these actions having been undertaken by management? The answer would appear to be: yes. This is a particularly important conclusion because the other factors which have been discussed are largely pre-determined: there is little individual managers can do about traditions of resistance, unemployment rates, the age and service of its workforce, the existence of unionized workers. Managers can, however, develop the type of strategies discussed above to reduce the possibility of resistance, and offset the unfavorable effects of some of the other factors. These strategies represent actions which, in these cases, were consciously formulated to eliminate, or at least reduce, the likelihood of overt confrontation over the closure. As such they represent the use of power, and this issue will be discussed in more detail in the following section.

MANAGERIAL STRATEGIES AS THE USE OF POWER

The strategies described in the preceding section represent the use of power by managers in their attempts to achieve a desired outcome which is, in this case, the implementation of a closure decision *in the absence of conflict*. The discussion of power has often been restricted to situations in which conflict already exists (for example, Dahl, 1957, 1961; Bachrach & Baratz, 1962, 1963, 1970; Polsby, 1963; Parsons, 1963; Wolfinger, 1971). Such work often assumes that conflict is a necessary pre-requisite to the exercise of power: without conflict, power is deemed un-necessary. Many definitions of power explicitly state this (for example, Pfeffer, 1981), while in other work the existence of conflict between actors is implicitly assumed.

Such power has been termed overt power (Hardy, 1982): it refers to the ability to produce preferred outcomes in the face of conflict between declared and active opponents. The sources of overt power are grounded in the differential access to material and structural resources which enable actors to influence decisions, nondecisions and the implementation of decisions. Power resources are conferred by dependency relations—because resources are scarce, some actors are dependent upon others for access to them (Emerson, 1962). Those who successfully possess and control these scarce resources are the powerful, able to influence decisions, agendas and resource allocations. Relevant power sources have been found to include the control of information, political access, expertise, assessed stature and credibility, the control of equipment, the control of rewards, punishments, referent symbols and legitimacy, the ability to cope with incertainty, prestige and status, (see, for example, Mechanic, 1962; Crozier, 1964; French & Raven, 1967; Hickson et al., 1971; Pettigrew, 1973; Pfeffer & Salancik, 1974; Pfeffer, 1981). These bases of overt power, if possessed and successfully mobilized (Pettigrew, 1973), enable actors to prevail in the face of competition.

Power can also be used to ensure that conflict does not occur: political actors may define success not so much in terms of winning in the face of confrontation (where there must always be a risk of losing), but in the ability to section off spheres of influence where domination is perceived as legitimate and thus unchallenged. The use of power in this situation revolves around attempts to create legitimacy and justification for certain arrangements so that they are never questioned by others. In this way power is mobilized not only to achieve physical outcomes but also to give these outcomes certain meanings: to legitimize and justify them.

> Political analysis must then proceed on two levels simultaneously. It must examine how political action gets some groups the tangible things they want from government and at the same time it must explore what these same actions mean to the mass public and how it is placated or aroused by them. In Himmelstrand's terms, political actions are both instrumental and expressive (Edelman, 1964:12).

Pfeffer has distinguished between substantive and sentiment outcomes of power. The former are physical outcomes which depend largely on resource-dependency considerations. The latter refer to the way people feel about these outcomes. Sentiments are mainly influenced by the use of political language, symbols and rituals.

If outcomes can be legitimized to the point where they are not questioned, even by potential opponents, actors have succeeded in obtaining their desired outcomes by using their power to *prevent* conflict from arising. This has been referred to by Lukes (1974) as the third dimension of power: where conflict is prevented by the shaping of preferences, perceptions and cognitions of potential opponents in such a way that they

> accept their role in the existing order of things, either because they can see or imagine no alternative, or because they see it as natural and unchangeable, or because they value it as divinely ordained and beneficial (p. 24).

Such processes depend heavily on the use of various symbols (Pfeffer, 1981) and myths (Cohen, 1975) to manage meaning (Pettigrew, 1977, 1979). These are relatively unexplored aspects of power, although the use of power to prevent conflict has recently been attracting some empirical attention (see, for example, Gaventa, 1980; Saunders, 1980).

This aspect of power has been termed unobtrusive (Hardy, 1982), not so much because power is used unobtrusively, but because of the circumstances in which its is used and the objective of its use. *Overt* power is employed in situations of overt confrontation, with the aim of *defeating* opposition. *Unobtrusive* power is used before overt confrontation occurs with the explicit aim of *preventing* it. The former will tend to rely on resource interdependencies, while the latter will usually employ more symbolic sources of power. It is not inconceivable, however, that resources such as expertise and information may be used to prevent opposition from arising, while symbolic power such as myths might be used to discredit active opponents. So, some power sources at least, can be employed both to prevent and defeat opposition. In fact, both types of power source are derived from the control of (more or less tangible) scarce resources.

Unobtrusive power is important for a number of reasons. First it incorporates an aspect of power that all too often has been ignored in the literature: the use of power in the absence of conflict. Second, it draws attention to the fact that actors may, if they feel threatened by the consequences of overt confrontation, undertake actions with the explicit objective of reducing the chances of this happening. Third, to assume that the absence of conflict automatically indicates some sort of "genuine" consensus or satisfaction is not theoretically justifiable because in some situations at least, quiescence may be the result of the use of power:

> We may, in other words, be duped, hoodwinked, coerced, cajoled or manipulated into political inactivity (Saunders, 1980:22).

The actions taken by the managers in the two case-studies can be described as the use of unobtrusive power: these actions were consciously taken with the explicit intention of reducing the likelihood of overt confrontation. In both cases managers perceived a threat of industrial action—they felt it to be a feasible possibility, and they were anxious to avoid it because of the negative consequences which could ensue (in terms of damaged reputations, difficulties in transferring production, or the setting of bad precedents).

These strategies are then, tied in with the perceived threats related to overt confrontation; where there is no perceived threat, such actions will be deemed unnecessary. Two other situations illustrate this point. In another closure in the U.K., Andersons were in a position where they did not anticipate any sympathy action from employees in other factories—it was such sympathy action that constituted a major threat in the Newlands closure. As a result they did not see the need to engage in the complex procedures discussed above; instead when the employees occupied the factory in a bid to stop the closure, they forced them back to work with the threat of reduced redundancy payments. In the Camerons closure discussed above, management did not perceive any risk of overt confrontation from the representatives of white collar staff because they were relatively nonunionized, and the unions were still negotiating for full recognition.

> At the time of the closure we'd only really just started to get properly unionized (staff representative).
>
> You couldn't ask union members to do anything—they'd run a mile (staff representative).

As a result no attempt was made to involve the staff representatives to the same extent as the shop stewards: although a staff steering group was set up, it was not until 1978 (two years later than its manual counterpart) and it lasted less than a year. Management simply did not feel the strategies that had been used vis-a-vis the manual workers were necessary for the white collar employees.

> The relative inactivity of staff unions in the closure process compared with the activity of the manual unions was affected by the timing of the closure relative to growth in staff unionization and the progressive involvement in negotiating rights. . . . . There was a lack of support from the staff themselves for their union to be involved at the late stage of 1978 when it was clear that redeployment and other problems by managers, including the redeployment manager, in a way that satisfied most staff. . . . . Works management's view is that the need for the establishment of a similar steering group for staff was doubtful (managerial report).

In these two examples, managers did not evoke the complex strategies described above because they did not feel threatened: at Camerons managers simply did not believe that resistance from staff employees would occur; at Andersons overt confrontation from these particular employees did not represent a threat because they believed it would be confined to the closing factory.

The actions in the two case-studies have been described as the use of unobtrusive power: they have been found to be consciously implemented with the express intention of avoiding conflict because of its negative consequences. A question still remains, however, as to whether these actions were instrumental in producing acquiescence. Naturally, we can not say that these strategies alone accounted for the absence of resistance: a number of other explanations have been found to be plausible. Nor can we say, however, that they had no effect at all. It seems likely that the chances of overt confrontation would have been higher had these actions not been taken and that, to a certain extent, they have had a significant bearing on the outcome. What is also important, regardless of whether they had any impact or not is the fact that management perceived a need for them, and that is why they were implemented. It also draws attention to the fact that if we are to fully understand the different responses to closure that occur we need to include the concept of unobtrusive power: to confine research on power to situations of conflict may explain some closures, but it ignores the vast majority which pass peacefully, some of which may be attributable to the type of strategies described here. To assume that all closures that occur without resistance are the result of employees being genuinely "happy" with their situation is probably misleading and unrealistic.

It is at this stage that we might return to the question of a definition of power. It has been argued that studies have often taken a relatively narrow definition of the term which results in an exclusive focus on overt power: broader definitions are necessary if we are to acknowledge the use of symbolic power. A definition which encompasses both aspects concerns the ability of an actor to structure the situation or perceptions of the situation such that others act as the actor desires. Power is thus conceived of as a generic term encompassing coercion, persuasion, influence and authority. The ability to structure situations refers to overt power, while the ability to structure perceptions describes unobtrusive power.

SUMMARY AND CONCLUSIONS

The above analysis focusses on two case-studies of closure in an attempt to explain why resistance did not occur. A number of explanations are presented, many of which had some impact on the eventual outcome. The focus, however, has been on the use of managerial strategies undertaken to prevent resistance, raising the issue of symbolic power to manage meaning.

The symbolic nature of the mechanisms used to legitimize the closure must be emphasized: the culture change at Camerons was instigated primarily for the benefit of the impeding closure; the new firm set up by Andersons saved only 200 jobs; the consultation in both cases did not involve the original decision to close; redundancy pay does not truly compensate for lost jobs, particularly if new employment is not forthcoming; the reasons given by Camerons were not the real reasons, and Andersons was prevented from proving its case because of confidentiality. The strategies were then, symbolic rather than substantive in the impact on employee/union perceptions of the closure.

The analysis also draws attention to the choice which faced these managers. If closure is a situation in which resistance can be expected: managers can opt to engage in the mobilization of their structural sources of power, should overt confrontation arise; or they can take steps to manage meaning and try to ensure that resistance does not occur. This choice is never completely free, however, as managers will have to take stock of their own resources. Within this constraint, however, it appears that in some situations managers perceive a choice and opt for unobtrusive strategies which do have some effect on the outcome.

The use of unobtrusive power to manage meaning raises a moral issue concerning the degree to which employees were manipulated. On one hand, symbolic power does involve offering some concessions:

> I was certainly much more liberal with the company's money than a manager in a normal situation, but I reckoned that it was my job to get it closed without any upset (Mountside works manager).

On the other hand, by influencing perceptions managers kept employees unaware of their potential power that might, if used successfully, have produced even greater benefits.

> What the stewards and officials wouldn't have appreciated is how worried the company was that the closure shouldn't go wrong. Because the company was worried if the officials had said: 'Fine, you close a works down but we're going to negotiate something special for that', the company would have negotiated so as to prevent trouble and the possibility of its spreading and affecting our business (Mountside works manager).

It should be said that to study certain actions is not to make any judgement on them and, if anything, framing these types of actions in a political analysis will do more to highlight, not obscure, the ethical questions.

Future studies of power should then, devote more attention to the use of power to prevent conflict, and the processes of symbolic power. This is not an easy task because we need to distinguish between

> Legitimacy which is voluntarily ceded from below, and various forms of manip-
> ulative relationships in which legitimacy is in some way imposed from above.
> The problem is that empirically, both types of power relationship are likely to
> appear the same (Saunders, 1980:28).

A starting point might be to focus on situations in which, although resistance might be expected, it does not occur. If dominant groups are prepared to admit that, as a result of perceiving a threat from overt confrontation, they took steps to try to prevent conflict, researchers may be able to explain acceptance by referring to these actions. Naturally, this scenario is fraught with problems: if research is retrospective, "admissions" may be more akin to post hoc rationalization; competing explanations will always exist, as is clear from these studies, and it may be difficult to allocate any precise degree of causality to them. These problems are excuses rather than reasons for not studying power: power has always been a difficult issue to study empirically, but confining ourselves to issues which are easily researchable is not the way to further knowledge.

Looking more broadly at these analyses of closure situations as examples of implementing strategic changes, this research has reaffirmed Quinn's (1980) general point that in the management of strategic changes there are critical process issues to consider as well as cognitive or intellectual considerations. Strategic changes do not just evolve and are implemented as a result of rational–logical debates, their process is shaped by disturbances in the firm's environment, by the speed and alacrity with which groups pushing for change can use such environmental disturbances to legitimate changes, and by the skill with which those advocating and implementing changes can draw on overt and unobtrusive power sources to openly defeat opposition or prevent opposition in a choice or change process.

NOTES

1.   The study (Hardy, 1982) also included two hospital closures, and the data from these helped to formulate the findings which are presented here.

2.   In Britain organizations are required to pay compensation for workers who lose their jobs through redundancy. The minimum is between one and one and a half week's pay per year of service, depending on how old the individual is. The state then reimburses 41% of this. Companies can pay more than this although they receive no additional reimbursement. In this study both companies paid between 2 and 3 times the state amount.

3.   Minimum notice of 3 months is required by law if more than 100 redundancies are involved.

REFERENCES

Allison, G. A., *Essence of Decision: Explaining the Cuban Missile Crisis*, Boston: Little Brown, 1971.

Andrews, K., *The Concept of Corporate Strategy*, Homewood, Illinois: Irwin, 1971.

Bachrach, P. & M. S. Baratz, "The Two Faces of Power," *American Political Science Review* 56 (1962), 947–52.

Bachrach, P. & M. S. Baratz, "Decisions and Nondecisions: An Analytical Framework," *American Political Science Review* 57 (1963), 641–51.

Bachrach, P. & M. S. Baratz, *Power and Poverty*, London: Oxford University Press, 1970.

Beynon, H., *Working for Ford*, London: Allen Lane, 1975.

Beynon, H. & H. Wainwright, *The Worker's Report on Vickers*, London: Pluto Press, 1979.

Boswell, J. S., *Business Policies in the Making*, London: Allen & Unwin, 1983.

Bower, J. L., *Managing the Resource Allocation Process*, Cambridge: Harvard University Press, 1970.

Bryer, R. A., T. J. Brignall & A. R. Maunders, *Accounting for British Steel*, London: Gower. (1982).

Carpenter, M., The Labour Movement in the NHS.

Chandler, A. J., *Strategy and Structure*, Cambridge: MIT Press, 1966.

Clarke, T., "Redundancy, Worker Resistance and the Community" in G. Craig, M. Mayo, N. Sharman (eds). *Jobs and Community Action*, London: Routledge & Kegan Paul, 1979.

Cohen, Anthony, *The Management of Myths*, Manchester: Manchester University Press, 1975.

Crozier, M., *The Bureaucratic Phenomenon*, Chicago University Press, 1964.

Dahl, R., "The Concept of Power," *Behavioural Science* 20 (1957), 201–15.

Dahl, R., *Who Governs*, London: Yale University Press, 1961.

Edelman, M., *The Symbolic Uses of Politics*, Urbana: University of Illinois, 1964.

Emerson, R. M., "Power-Dependence Relations," *American Sociological Review* 27 (1962), 31–41.

French, J. R. P. & B. Raven, "The Basis of Social Power," in D. Cartwright & A. Zander (eds.) *Group Dynamics*, New York: Harper & Row, 1968.

Galbraith, J. R. & D. A. Nathanson, *Strategy Implementation: The Role of Structure and Process*, St. Paul: West Publishing, 1978.

Gaventa, J., *Power and Powerlessness*, London: Oxford University Press, 1980.

Greenwood, J. *Worker Sit-ins and Job Protection*, London: Gower Press, 1977.

Hardy, C., *Organisational Closure: A Political Perspective*, Ph.D. Thesis, Warwick University, 1982.

Hemingway, J. & W. Keyser, *Who's in Charge?*, Oxford: Metra, 1975.

Hickson, D. J., C. R. Hinings, C. A., Lee, R. E. Schneck, & J. M. Pennings, "A Strategic Contingencies Theory of Intra-Organisational Power," *Administrative Science Quarterly* 16 (1971), 216–29.

Hofer, C. W. and D. E. Schendel, *Strategy Formulation: Analytical Concepts,* St. Pauls: West Publishing, 1978.

Hyman, R. & R. H. Fryer, "Trade Unions—Socialogical and Political Economy" in J. B. McKinlay (ed.) *Processing People,* London: Holt, Rinehart, & Winston, 1975.

Kanter, R. M., *The Change Masters: Innovation for Productivity in the American Corporation,* New York: Simon & Schuster, 1983.

Lane, T., "We're Talking about a Closure Movement" in Rainnie & Stirling (eds.) *Plant Closure,* Newcastle: Faculty of Community and Social Studies, Newcastle-Upon-Tyne Polytechnic, Occasional Paper, 1981.

Luce, S. R., *Retrencment & Beyond: The Acid Test of Human Resource Management,* Ottawa: The Conference Board of Canada, 1983.

Lukes, S., *Power: A Radical View,* London: MacMillan, 1974.

March, J. G. & J. P. Olsen, *Ambiguity and Choice in Organizations,* Bergen: Universities, for Laget 1976.

Matthews, J., *Ford Strike,* London: Panther, 1972.

Mechanic, D., "Sources of Power of Lower Participants in Complex Organisations," *Administrative Science Quarterly* 7 (1962), 349–64.

Miller, D. & P. Friesen, Momentum and Revolution in Organizational Adaption, *Academy of Management Journal* 23 (1980), 591–614.

Mintzberg, H., Patterns in Strategy Formation, *Management Science* 24: 9 (1978), 934–948.

Mintzberg, H. & J. A. Waters, Tracking Strategy in an Entrepreneurial Firm. *Academy of Management Journal* 25: (3) (1982), 465–99.

Mumford, E. & A. M. Pettigrew, *Implementing Strategic Decisions,* London: Longman, 1975.

Parsons, T., "On the Concept of Influence," *Public Opinion Quarterly* 27 (1963), 37–62.

Pettigrew, A. M., *The Politics of Organisational Decision-Making,* London: Tavistock, 1973.

Pettigrew, A. M., Towards a Political Theory of Organisational Intervention *Human Relations* 28: 3 (1975), 191–208.

Pettigrew, A. M., *The Creation of Organisional Cultures,* Brussels: European Institute for Advanced Studies in Management, Working Paper 7–11, 1977.

Pettigrew, A. M., "On Studying Organisational Cultures," *Administrative Science Quarterly* 24 (1979), 570–81.

Pettigrew, A. M., The Politics of Organizational Change in Niels Bjorn Anderson (ed). *The Human Side of Information Processing,* Amsterdam: North-Holland, 1980.

Pettigrew, A. M., *Context and Politics in Organizational Change,* Englewood Cliffs: Prentice Hall and Oxford, Basil Blackwell. 1984.

Pfeffer, J., *Power in Organisations,* Springfield, Mass: Pitman, 1981.

Pfeffer, J. & G. Salancik, "Organisational Decision-Making as a Political Process," *Administrative Science Quarterly* 19 (1974), 135–151.

Pizzorno, A., "Political Exchange and Collective Identity in Industrial Conflict" in C. Crouch & A. Pizzorno (eds). *The Resurgence of Class Conflict,* London: MacMillan, 1978.

Polsby, N. W., *Community Power and Political Theory*, London: MacMillan, 1963.

Porter, M., *Competitive Strategy*, New York: Free Press, 1980.

Quinn, J. B., *Strategies for Change; Logical Incrementalism*, Homewood, Illinois: Irwin, 1980.

Quinn, J. B., Managing Strategies Incrementally *Omega* 10: 6, (1982), 613–627.

Ranson, S., R. Hinings, & R. Greenwood, "The Structuring of Organisational Structure," *Administrative Science Quarterly* 25: 1 (1980) 1–14.

Red Notes Pamphlet *Fighting the Lay-Offs at Ford*, London: Red Notes, 1978.

Saunders, P., *Urban Politics*, London: Penguin (1980).

Ursell, G. D. M., *A Sociological Study of the Containment of an Extra-Union Organisation Amongst Steel Workers*, MSc: Bradford, 1976.

Vulliamy, D. & R. Moore, *Whitleyism and Health*, London: Workers Education Association, 1979.

Whetten, D. A., "Sources, Responses, and Effects of Organisational Decline," in J. Kimberly & R. Miles (eds). *The Organisational Life Cycle*, San Francisco: Jossey-Boss, 1980.

White, P. J., "The Management of Redundancy," *Industrial Relations Journal* 14 (1983), 32–40.

Wolfinger, R. E., "Nondecisions and the Study of Local Politics," *American Political Science Review* 65 (1971), 1063–80.

Wood, H. S. & J. Cohen, "Approaches to the Study of Redundancy," *Industrial Relations Journal* 8: 4 (1977/8), 17–19.

INDUSTRIAL RESEARCH IN THE AGE OF BIG SCIENCE

Margaret B. W. Graham

Historians have observed that one of the most important features of twentieth century industrial life in the United States has been the ability of large corporations to harness technology as an industrial resource. To assemble the integrated system of industrial research and development that has allowed the corporation to develop and apply technology at will, has been no small achievement. Moreover, science-based research has been an atypical managerial activity. The people and the institutions that have performed industrial research have often been motivated by values that have conflicted with those of their companies, and research itself has differed from more standard management functions in the matters of timing, and the need for stability and degree of uncertainty.

For much of the twentieth century the hub of the industrial research

Research on Technological Innovation, Management and Policy
Volume 2, pages 47–79
Copyright © 1985 by JAI Press Inc.
All rights of reproduction in any form reserved.
ISBN: 0-89232-426-0

activity inside the corporation has been the corporate research center.[1] This paper concerns the evolution of the corporate research center's relationship to the corporation at large, the changes it underwent at the midpoint of the twentieth century and the effect these changes had on the way research was conducted.

World War II shifted the way research was conducted in industry, a shift that coincided with what is termed the Age of Big Science. Two changes contributed to the shift: a change in the role of science in American society ("the research climate") and a change in the way research was funded and organized in individual companies ("the industrial research context"). The shift magnified the importance of research in industry and altered prevailing notions concerning the appropriate use of science in industry.

INSTITUTIONAL STEREOTYPES

Before we turn to the historical causes of the shifts in industrial research climate and context, we can compare the stereotypical industrial research laboratory before the Second World War to its counterpart in the Age of Big Science. In his book *America by Design* historian David Noble characterized the typical twentieth century relationship of the research enterprise to the rest of the corporation as "collective subservience." Research management under the system that existed at the time resembled the manufacture of ideas. To quote Noble further:

> As the industrial research laboratories grew in size, the role of the scientists within them came more and more to resemble that of the workmen on the production line and science became essentially a management problem. The industrial laboratory was quite different from its university counterpart, which supplied it with scientific personnel. Whereas the university researcher was relatively free to chart his own paths and define his own problems (however meager the resources), the industrial researcher was more commonly a soldier under management command, participating with others in a collective attack on scientific truth.[2]

Although Noble's portrayal of industrial research laboratories in the early twentieth century overlooks important individual achievements by some noteworthy industrial researchers it highlights several characteristics of the climate for industrial research that concern us here. In content and approach there was a clear distinction between industrial research and university research. University laboratories were oriented primarily towards instruction and their research mission concentrated on fundamentals. They were supported, seldom generously, either by internal university funds or by foundations. Their researchers had to be willing to give up research time for teaching, to accept minimal pay, and to forego expensive equipment for the privilege of defining the problems on which they chose to work.

By contrast, technical staff members of industrial laboratories received better pay and enjoyed ample research support. In return, they were likely to be told what problems to work on and how to conduct their research. In the majority of cases the work would be clearly linked to the product or process needs of the companies that employed them. Industrial researchers who did not share the objectives of the organizations for which they worked could not expect to remain with their companies for very long.

Seldom did these two different types of research organization compete with each other. Indeed, close cooperative relationships frequently existed between university and industrial research laboratories, involving a straightforward division of research turf.[3]

By contrast, R & D management literature from the 1950s and 1960s conveys a picture of research in a corporate setting that differed sharply from that of the 1920s. Popular topics for discussion among laboratory managers in the conference sessions were the psychology of the scientist, researcher motivation, and research incentives. The stereotypical corporate research center of the latter era had become a corporate counterculture, inhabited by independent and rather impractical geniuses who had to be coaxed to participate in projects that were of immediate utility to the company. In the Age of Big Science the prewar arrangement of clear-cut distinctions and a logical division of labor between university and industrial research organizations had given way to one in which the demarcation was much less tidy, and competition had replaced cooperation.

It should be noted that the shift described here was particularly true for the corporate research organization in the electrical, later electronics industry where the discipline in question changed from chemistry to physics.

INDUSTRIAL RESEARCH BACKGROUND

Organized research as a major corporate activity conducted inside corporations began in the late nineteenth century. By the World War I a handful of large and progressive companies such as General Electric, American Telephone and Telegraph, DuPont and Standard Oil of Indiana, all had laboratories that were intended to pursue long-term technical advantage for their companies. Recent works by historians of business and technology have shown that these companies and others used industrial research to broaden their product options, to consolidate their competitive dominance of key industries through patent control, and to improve their cost positions through leadership in process development.[4] The place of industrial research in the United States economy expanded still further during World War I, stimulated by such factors as clearly defined national goals, a temporary government moratorium on patent control, an emergency need for substitute materials, and a sudden infusion of government funding.[5]

So impressive were the benefits that some companies, mostly in the chemically based industries, gained from their wartime research activities, that many new companies set up research facilities of their own after the war. As early National Research Council surveys of industrial research laboratories found, by 1921 more than 350 companies claimed to have some type of dedicated R & D facilities employing more than 1600 scientific and technical personnel.[6] Only a few of these new facilities engaged in research of any scientific consequence. These were principally companies that hoped to emulate DuPont or General Electric in finding major new categories of products that would enable them to redeploy excess productive capacity built up during the war. Most of the so-called research laboratories were actually used for the humbler purposes of product testing and market support. Nevertheless, the proliferation of such facilities did create a market for industrial researchers, trained engineers and scientists who were willing to pursue industrial problems and applications in an industrial setting.

Research between the wars became a much more publicly visible form of industrial enterprise, an activity that could be blamed as well as praised for its effect on the national economy. Companies discovered that social, and therefore political, attitudes towards industrial research could change quickly, and in many companies this perception on the part of the management affected the level of support for research as much as did research output.

During the 1920s industrial research received credit in the press for enormous improvements in industrial processes and productivity gains; but these same contributions caused research to be vilified in the 1930s when public attention focused instead on the way technology had caused market saturation and consequently a reduction in the number of jobs needed.[7]

Hundreds of researchers lost their jobs in the early 1930s as 300 of the 1500 laboratories that had been opened by 1930 closed down altogether. Others reduced the size of their technical staffs. The cuts reflected more than corporate attempts to reduce non-essential overhead expenses, for there was a widespread view that public backlash was justified. It seemed to some corporate executives that numerous editorials and articles might be right when they called for a halt to further research. It was said that humanistic studies would have to catch up with advances in scientific discovery if ways were to be found to incorporate the fruits of scientific research into constructive channels.[8]

To counter the damaging rollercoaster phenomenon of public opinion, American scientific leaders conceived of a more active and permanent role for the government in the shaping of the national climate for research. Research universities had by the 1930s developed an interest in the preservation of jobs for their graduates and their spokesmen concluded that a major effort was needed to coordinate and focus scientific activity. The public had to be educated, they believed, to see the creative rather than the destructive possibilities of research. To this end, President Franklin Roosevelt, in 1934, invited a group of prominent physicists to form the Science Advisory Board. The Board was headed by Karl Compton, the newly-chosen president of the Massachusetts Institute of Technology. Compton argued that the problem was not inherently a scientific one, but rather a "misuse of science" on the part of business. It should be the job of science, he wrote in a letter to the *New York Times,* to "find ways of using the fruits of overproduction to create new industries."[9] In its advocacy of a more active government role the Science Advisory Board broke with the longstanding opposition of the American scientific community to all forms of government intervention, opposition born out of fear of political interference with academic freedom.[10] The Board's report called for a National Science Board to be established similar to Great Britain's arrangement for supporting fundamental scientific research. The report further amplified Compton's charge that industry had "misused science" by allocating its development activities too much towards short-term purposes that eliminated jobs rather than creating new products and new industries.[11]

Further to influence the terms of public debate on a more sustained basis, several prominent physicists and leaders of industrial research (including Compton of MIT and Frank Jewett of American Telephone and Telegraph) banded together to form a second forum for the public promotion of science, the Advisory Council on Applied Physics. Their efforts to gain recognition for the need for scientific research in industry had the desired effect. By the year 1937–38 more than fifty companies were featuring their research accomplishments prominently in their annual reports. Westinghouse, for example, which had long supported applied research, set up a new program to promote fundamental research within its technical community. Each year ten people were selected to receive two-year Westinghouse Research Fellowships carrying a stipend of $2400 per year. Fellows were encouraged to work on projects that interested them but were not of immediate interest to the Westinghouse Laboratory.[12] Thirty to forty qualified applicants sought the chance each year to be one of the ten people on a 300-person technical staff to work on problems of their own choosing.

Academic scientists blamed industry's "misuse of science" in part on the Federal Government's policy. Having stimulated research and development activity in chemistry and radio during World War I, the Federal Government had thereafter left the field to private industry. The result had been a further concentration of patents, and the research facilities that produced them, under the control of a few major companies, often only one company in an industry. Critics charged that these dominant companies had been permitted to control the evolution of their respective industrial technologies in their own narrow interests, while keeping smaller or newer competitors from engaging profitably in research.[13] Any attempt to address the problems of technical dominance by a few companies would entail a significant change in the structure of the organization of research in the U.S. economy. In particular, a formal peacetime relationship would have to be forged among the several independent institutions—research-oriented universities, private research institutes, technology-based companies, and government—that together formed the climate for scientific research in the United States, and that had traditionally cooperated only during wartime.

The scientific community deferred its agitation for peacetime reform of science policy when World War II imposed its own pressing demands for coordination of scientific research. But the leaders of the scientific community would not forget the lessons they had derived from their interwar experience. They had learned that what they regarded as appropriate uses of science were often in conflict with typical corporate interests. During the war they found staunch allies among politicans who held compatible views. Chief among them was Senator Harley Kilgore of West Virginia who wanted to coordinate scientific activity and R & D funding beyond the standard established group of companies by setting up an Office of Technology Mobilization. Nothing came of the senator's proposal during the war, but Kilgore and his group of liberal sympathizers formed one of the major parties in the National Science Foundation debate that went on for five years after the war ended.

THE ERA OF BIG SCIENCE

World War II brought an enormous increase in federal research funding. Funding for industrial R & D rose tenfold, from $48 million in 1940 to $500 million in 1945. Funds were distributed through such channels as the individual military procurement offices and the Office of Scientific Research and Development.[14] Increased federal intervention had an enduring effect on the performance of industrial research in general, but the effect was particularly strong in technology based industries such as aircraft and electronics which were in the early phases of their development. First, more companies in more industries became active in performing R & D. Research took on a new competitive dimension. Second, many companies reorganized their research communities, centralizing all or part of previously decentralized facilities and often separating research from manufacturing facilities. Third, decisive changes occurred in the composition of technical staffs, though the direction of the changes varied depending on the type of research a company chose to emphasize.

Unlike the end of World War I, the conclusion of World War II brought only a brief pause in direct research funding by government. Scientific leaders argued that it was a matter of national interest that the momentum and continuity built up on wartime projects such as the atom bomb, radar and jet propulsion should not be allowed to subside, and that funding begun under wartime conditions should continue. Supporters of a strong national scientific policy argued that effective defense preparedness depended on scientific leadership. Moreover, commercial application of the technologies developed for military purposes seemed to offer unlimited opportunity.[15]

Business leaders by contrast, were skeptical of government largesse towards research. Big business vehemently opposed early post-war proposals by Senator Kilgore to establish a peacetime scientific mobilization agency. Two issues were paramount—the ownership of patents stemming from government funded work, and whether agency control should reside within the scientific community itself or in the executive branch of government. It was 1950 before prolonged national debate finally ended in the compromise measure that created the National Science Foundation. Business opposition was only defused by limiting the agency's role solely to funding fundamental research within universities, control to be vested in an executive appointee advised by a strong independent committee of scientists.[16]

While the proposal for a National Science Foundation received a great deal of public attention, growing amounts of government money, which poured into research in industry through other channels, attracted little notice. Defense-related agencies such as the Office of Naval Research, the Atomic Energy Commission, and later the National Aeronautics and Space Administration allocated massive and growing amounts of federal money to industrial research and development (Tables 1 and 2). The figure spent by the Defense Department alone was $600 million in 1948, rising to the peak of over $7.5 billion in 1966. Of this amount roughly a quarter went to research in its several forms (from purely basic to wholly applied), and the rest to development. Basic research never surpassed seven percent of the total, but it played a more important role in operational terms than the absolute amounts invested would have indicated.

Most of the scientific benefits of wartime research funding during World War I had accrued to the chemical based industries. This time direct funding concentrated overwhelmingly in three other industries that had reached a stage of technological evolution similar to that of the chemical based industries after World War I. These were aircraft, electrical goods, and instruments. In the aircraft industry only twenty percent of the money that supported research came from private funds. In the other industries roughly half of the R & D support came from the companies themselves, but the lion's share of fundamental research funding was provided by the government (Tables 3 & 4).

The increased availability of government funds to support research prompted numerous institutions and companies either to set up new research facilities or to broaden the scope of their research activities. More companies began to conduct the fundamental research previously

Table 1. Federal Expenditures for R and D and R and D Plant, by Agency, Fiscal Years 1940–1966 (millions of dollars)

Department or Agency	1940	1948	1956	1964	1966
Agriculture	29.1	42.4	87.7	183.4	257.7
Commerce	3.3	8.2	20.4	84.5	93.0
Defense	26.4	492.2	2,639.0	7,517.0	6,880.7
Army	3.8	116.4	702.4	1,413.6	1,452.1
Navy	13.9	287.5	635.8	1.724.2	1,540.0
Air Force	8.7	188.3	1,278.9	3.951.1	3.384.4
Defense agencies	—	—	—	406.9	464.5
Department-wide funds	—	—	21.9	21.1	39.7
Health, Education and Welfare	2.8	22.8	86.2	793.4	963.9
Interior	7.9	31.4	35.7	102.0	138.7
Atomic Energy Commission	—	107.5	474.0	1,505.0	1,559.7
Federal Aviation Agency	—	—	—	74.0	73.4
National Aeronautics and Space Administration	2.2	37.5	71.1	4,171.0	5,100.0
National Science Foundation	—	—	15.4	189.8	258.7
Office of Scientific Research and Development	—	0.9	—	—	—
Veterans Administration	—	—	6.1	34.1	45.9
All other Agencies	2.4	11.8	10.4	39.7	66.1

Source: Federal Funds for Science XIV (National Science Foundation, 1965).

regarded as a university preserve. Partly this was a response to the prevailing notion that the United States could no longer afford to depend on foreign sources for the basic science that fed its industry, and partly it stemmed from the view that industry had too long been a parasite on academia where research was concerned.[17] At the same time universities took on more applied work, building on wartime projects. Johns Hopkins University for example in its Applied Physics Laboratory, extended its wartime work on the proximity fuse into research on guided missiles.[18] The new overlap between the types of research conducted in industry and in academia resulted in direct competition for technical and scientific personnel between the two sectors.

Nationwide the mix of technical personnel employed by industrial research organizations shifted in the 1940s and the 1950s, continuing in directions established in the late 1930s. Increased demand for research trained personnel coupled with rising education levels brought about a shift toward younger and more theoretically trained research-

Table 2. R and D Performance, By Sector

Sources of R & D Funds (Sector)	Federal Government	Industry	Colleges and Universities	Other Nonprofit Organizations	Total
1953 Transfer of funds ($ millions)					
Federal Government	1,010	1,430	260	60	2,760
Industry		2,200	20	20	2,240
Colleges and Universities			120	—	120
Other nonprofit organizations			20	20	40
Total	1,010	3,630	420	100	5,160
1963 Transfer of funds ($ millions)					
Federal government	2,400	7,340	1,300	300	11,340
Industry		5,380	65	120	4,565
Colleges and Universities			260	—	260
Other nonprofit organizations			75	110	185
Total	2,400	12,720	1,700	530	17,350

Source: National Science Foundation Review of Data on Science Resources, Vol. I, No. 4, May 1965.

56

Table 3. Funding of R and D, By Source

Industry	Total Cost of Industrial Research and Development (in millions)	Cost of Federal Government-Financed Research and Development	
		Total Cost (in millions)	Percent of Total Cost
All industries..........	$3,664.4	$1,357.9	37.1
Aircraft and parts	758.0	639.8	84.4
Electrical equipment.........	743.3	404.0	54.5
Professional and scientific instruments	171.7	76.8	44.7
Telecommunications and broadcasting.........	113.0	58.9	52.2
Machinery	318.9	57.2	17.9
Fabricated metal products and ordinanace.....	103.3	32.7	31.6
Chemicals and allied products	361.1	8.9	2.5
Petroleum products and extraction.........	145.9	8.2	5.6
Primary metal industries.........	59.8	4.5	7.5
Other industries..........	889.4	66.2	7.4

Source: National Science Foundation.

57

Table 4.   Principal Industrial Contractors with the Office of
Scientific Research and Development during World War II

	Contracts	Total Funding
Western Electric Company	94	$17,091,819.00
Research Construction Company	2	13,950,000.00
General Electric Company	58	8,077,047.14
Radio Corporation of America	54	5,783,498.13
E. I. du Pont de Nemours and Company	59	5,704,146.54
Westinghouse Electric Manufacturing Corporation	54	5,122,722.26
Remington Rand, Incorporated	3	4,654,050.00
Eastman Kodak Company	29	4,509,200.00
Monsanto Chemical Company	8	4,222,044.00
Zenith Radio Corporation	3	4,175,000.00

Source: James Phinney Baxter 3rd, *Scientists Against Time* (Boston: Atlantic-Little Brown, 1946), pp. 456–457.

ers. In 1940, industry employed 145,000 scientists of which 27.5 percent held PhDs compared to 300,000 engineers of whom 5 percent held PhDs. By 1950 the scientists numbered 245,000 of whom 43.5 percent held PhDs, while engineers had increased to 545,000 of whom 15 percent were PhD holders.[19] PhD engineers and advanced degree holders in chemistry and physics were in great demand, with PhD physicists hardest to hire. The aggregate numbers hide differences among research based industries. The older technology based companies, having reached more stable phases in their technologies tended to concentrate on applied work and technology transfer activities. In the newer technologies, such as electronics, credential inflation was especially problematic. This was reflected in the relative prosperity of physics as a discipline. By 1962, the average salary of a physics researcher in the United States was $13,000, $1,000 more than the average chemist's salary and several thousand dollars higher than the average for all fields. At the most senior levels, the differential widened. Physics PhDs commanded an average of $18,000 versus $16,000 for PhD chemists at the same levels.

The new breed of researchers, especially the physicists, saw themselves as part of a fraternity that had attained the highest status. They were mobile; they were in increasing demand; and their knowledge was transferrable among a growing set of companies. They also had in common values that reflected a change in what might be termed the national research philosophy. Among these values was the view that loyalty to the advance of a scientific discipline was a worthier personal goal than loyalty to any institution, company or university.

To many intellectuals this professional identification served the higher needs of American society. Talcott Parsons, renowned Harvard sociologist, summed up the new credo of many American intellectuals in 1958, when he called the New Deal the end of a practical era in the United States. Parsons maintained that in the coming era the producers of the ideas would also become the initiators, while businessmen would be relegated to the less imaginative activity of administration. "The business executive class," he predicted, "becomes the implementer in innovations which originate in the research process."[20] The emphasis on the primacy of ideas gave rise to a strong preference for theory over empiricism. Harvard's president, James Conant, maintained in his influential book *Science and Common Sense* that the degree of scientific understanding involved in industrial activity reduced the amount of risk involved in new technical enterprises. The higher the degree of knowledge concerning scientific fundamentals and the lower the degree of empiricism, he claimed, the more chance a project would have of successful commercial outcome.[21]

Novice industrial researchers of the post-war era, motivated by the new research philosophy, adopted a research style that was both theoretically motivated and capital intensive. Rejecting the old "cut and try" method that had involved large teams of experimenters, they regarded incremental improvement of existing practice as anti-intellectual, "brute force," and unworthy of rigorous scientific training. Moreover, researchers expected to have expensive capital equipment to prove out their theories, and numbers of technicians to assist them. The major wartime projects, after all, had spared no expense in research equipment and technical support.

To summarize, both the national research climate and the industrial research context changed dramatically in the Age of Big Science. Science moved out of its academic isolation and its supporting role in industry to become a prominent and aggressive initiator of technology-based innovation. Factors in the climatic change were a massive infusion of Federal funding for industrial research in certain focal industries and a new competition for funds and researchers among different research-performing institutions. Concurrently, the industrial context for research changed to accommodate a more diversified research mission. It absorbed a pool of research personnel who were committed to both a research philosophy and a set of values that were at odds with conventional corporate values. Management of research in such altered circumstances was a different kind of pursuit from what it had been before the war. The new breed of researchers would neither be harnessed nor directed, as their predecessors had been. Rather their ideas had to be contained and channeled with the hope that an occasional breakthrough discovery of immense value would compensate for much that was not needed or not useful in the commercial context.

All industrial research laboratories were affected in some manner by the shifts in the national climate for research but the degree to which their research activities were affected differed. While the decisive factor was the stage to which an entire industry's technology had developed, other influences on the character of an individual company's research was the position of an individual company in its industry and the company's structure. To see how corporate research organizations could respond to the changes described we will examine the experience of a leading corporate research laboratory in the electronics industry, the David Sarnoff Research Center of the Radio Corporation of America, located in Princeton, New Jersey.[22]

THE RCA EXPERIENCE

The Radio Corporation of America (RCA) experienced a radical transformation of its research function in response to wartime changes in the national climate for research. The evolving role of the corporate research laboratory inside RCA's corporate structure mirrored patterns of institutional behaviour that were present in some form in many other companies during the era of Big Science. In the decades following World War II, RCA's corporate research center changed from a support function to a major determining influence on the company's strategy.

Technology was at the heart of RCA's corporate identity. The Radio Corporation orginated as a government sanctioned technical monopoly, established in 1919 to serve in part as a repository for the combined radio-related patent holdings of General Electric (GE), Westinghouse, American Telephone and Telegraph (AT&T) and others. For the first few years of the company's existence the research organizations that had produced the patents remained with their parent companies; but in 1930

RCA took direct control of some of the GE and Westinghouse research facilities when it absorbed the manufacturing plants to which they were attached. To these RCA added a valuable research group that had been part of the old Victor Manufacturing organization that the company had acquired in 1929.

RCA's president David Sarnoff gained a reputation in the 1930s for a steadfast commitment to research. His attitude earned him the personal loyalty verging on devotion of RCA's research staff who were quite aware that many of their counterparts in other companies had lost their jobs in the economic downturn. In 1934 Sarnoff went further by making research a corporate expense and giving the company's core technical effort its first real measure of independence from the manufacturing operations. The money thus allocated, $880,000 in the first year, of which $740,000 went to manufacturing research and the balance to communications research, could legitimately be regarded as research revenue. It came from royalties paid to RCA by the many licensees who used its radio patents. In effect, RCA collected fees which it then used to conduct research and development activities on behalf of the entire radio industry. Because in this way it would determine which technologies were pursued and how they should be configured, it was able to maintain a substantial technological lead on the rest of its industry.

In the 1930s RCA's research activities were dedicated to advanced product development and new business development. Those who directed RCA's research functions wanted the company to invest also in what they termed "general" (i.e., unprogrammed or long-term) research, as a few outstanding research centers such as those of GE, Westinghouse, and AT&T were doing at the time. But David Sarnoff decided how the research budget would be allocated and for the time being he chose to concentrate all his company's discretionary research funding on RCA's major new product opportunity at the time, which was television.

WARTIME AND POSTWAR RESEARCH

When in 1940 the Federal Government first began to direct major new amounts of funding to radio-related research in conjunction with the Lend-Lease Plan, RCA took the important step of creating a central research facility. At first it was central both in its location and in its responsibility. It was located in Princeton, New Jersey, a site that was convenient both to the company's major manufacturing locations and to the corporate headquarters in New York City. Even before the building opened, the research organization that it was to house became a separate division, responsible for "all research, original development and patent and licensing activities of the corporation."[23] Control over RCA's entire research budget extended to some original development activities that were located at RCA manufacturing sites.

The new corporate research center in Princeton provided RCA with the chance to combine under one roof, and to conduct in absolute secrecy, the contract research that RCA had agreed to perform for the Government. It also allowed the company to move more in the direction of the "general" research that its research directors had been advocating for some time. For the duration of the war classified government work accounted for most of RCA's research activity. The company ranked fourth among leading industrial contractors with the Office of Scientific Research and Development, its fifty-four contracts totaling nearly $6 million.

After the war RCA's research and development activities both expanded and spread out, becoming ever more important in the eyes of the company's top management. A driving concern for the corporate laboratories' management after the war was to keep together through the transition from wartime to peacetime production the enlarged and highly specialized research staff that it had assembled. It pushed for immediate investment in research and development for color television along with other product opportunities that would use research skills akin to those that had been needed for defense research. In the case of color, which required some very advanced work on picture tubes, the timing of product introduction in the early 1950s was motivated less by market readiness than by the post-war needs of the research center to keep its staff fully employed. By 1950, the number of people listed as part of the total technical staff at RCA was 2723 people, over three times the size of the pre-war technical staff. Of these, 469 were employed at the Princeton research center, the rest at various manufacturing locations.[24] In the 1950s these manufacturing divisions dispersed geographically when RCA relocated some of its main manufacturing sites.

RCA became known inside and outside the company as a corporation dominated by its technical community. The Princeton research center was elevated in symbolic terms soon after the war, when in

1951, David Sarnoff renamed the facility the David Sarnoff Research Center. As further evidence of his favor, the Laboratories building added a suite of rooms for Sarnoff to use when he wished to entertain the Laboratory managers and honored guests on his frequent trips to Princeton. To be a senior manager at the Laboratories during this era proved to be a route to corporate advancement. Elmer Engstrom moved from heading the Laboratories to becoming first, head of a manufacturing division, and later President, then Chief Executive Officer, of RCA. George Brown, who directed important parts of RCA's color television development project in the late 1940's and early 1950's at the Laboratories, later became head of Research and Engineering, head of Licensing and a member of the company's Board of Directors. James Hillier, who headed the Laboratories in the late 1950's and early 1960's, also later became head of Research and Engineering in New York. Bolstered by chief executive commitment and the backing of other powerful representatives in New York, the corporate research center was in a position to exert considerable influence on top-level decision making at RCA well into the 1960s.

THE CHANGING ROLE OF RCA RESEARCH

Soon after the war the RCA Research Center began to move towards defining a long-term research mission for itself, rather than simply responding to the intermediate term product development needs of the separate product divisions. The Research Center could take initiative in this direction for several reasons. Electronics technology was at a point where much fundamental exploratory work needed to be done if its commercial potential were to be fully exploited. Further, David Sarnoff himself believed in the idea of self-obsolescence, the kind of revolutionary thinking that enabled a company to apply science to produce major technological leaps. In a more practical vein, RCA's research community enjoyed a high degree of financial autonomy. Even though RCA's research budget had expanded many times over its pre-war level, the corporate research center was able to remain completely independent of funding from the manufacturing divisions, a state that continued until the mid 1950s. In addition to royalties from the various domestic licensees, among whom were counted the various RCA product divisions, government contract work accounted for roughly twenty-five percent of the research budget.

By the early 1950s the David Sarnoff Research Center's explicit research mission was to conduct what was termed "building block research," i.e., filling in the knowledge base needed to support advances in semiconductor technology. Physics and Chemistry headed the list of activities that the Princeton research center reported to the National Research Council's survey of U.S. research laboratories in 1950. The brief description went on to say that the Princeton facility did fundamental research for the companies and divisions of RCA and its licensees, while the technical staffs at the manufacturing divisions did applied work on electronic devices, new products, new equipment, and new production processes. In fact, a large amount of problem-solving work continued to be done at Princeton, in support of such products as color television, video-recording, computers, and electron beam microscopy, to name just a few. But research, some of it of a truly fundamental nature, had become the preeminent activity, and had a pervasive influence on the organization as a whole.

Post war changes in the composition of the RCA research staff paralleled those in other research-oriented companies that were in the early stages of developing new technologies. Having hired a number of scientific specialists during the war, RCA's corporate research center had a research staff the equal of any in the nation in the areas of vacuum tube design, high frequency techniques, electron optics, acoustics and luminescent materials. In 1950, the senior technical staff at Princeton included 123 engineers, 53 physicists and 14 chemists, the scientific categories having grown more rapidly than the engineering ones. Even members of the technical staff who had worked for RCA before the war had changed their ideas about the way research should be conducted, having spent months or in some cases years, on loan to any of the several major wartime projects. One RCA engineer whose formal education had ended with his B.S. in electrical engineering at a state university later described the experience of working at the "Rad Lab" (the radar research project at the Massachusetts Institute of Technology), as one of the most exciting experiences of his life. It had been, he noted, an opportunity to work with some of the most talented research scientists and engineers of his generation.

RCA found itself competing after the war with many other electronics research organizations in its effort to hire research talent. The number of research establishments in electronics alone had expanded to 200 from just under thirty before the war. The supply of trained researchers was terribly limited, and qualified people had numerous offers from which to choose.[25] Elmer Engstrom in his position as head of Research and Engineering for RCA now regarded basic research as vital for RCA to pursue if the company were to attract its share of MA and PhD candidates from leading research universities. All RCA recruits would not, of course, be engaged in research into fundamentals, but the presence of such research at the Laboratories would create the type of intellectually stimulating environment that researchers were used to having in the universities from which they came. Throughout the decade of the 1950s Engstrom spoke out in vigorous opposition whenever it was suggested, often by spokesmen for the university scientific community, that basic research should be confined to the campus.[26]

The RCA research center held a number of attractions for prospective recruits: a campus-like environment in close proximity to Princeton University, a technical support staff that numbered two per research professional (double the ratio before the war); and much more sophisticated research equipment, mainly financed by the government. The Princeton facility's similarities to a university extended to its methods of choosing research projects and allocating personnel among them. Such decisions were negotiated among the laboratory directors, collegially assembled. New researchers had a considerable degree of latitude in selecting the projects on which they wished to work, and there was an unusual amount of opportunity for them to do exploratory work on an individual basis, also termed "sustaining research."

The arrival of large numbers of discipline-oriented researchers had a pronounced effect on the Princeton Laboratories culture. Already there existed a split between the staff that had come from RCA's consumer product facility in Camden (the former Victor manufacturing plant) and the device-oriented research group that had been doing advanced vacuum tube development at the former General Electric plant in Harrison, New Jersey. The newcomers with their academic preferences and expectations and their complete lack of manufacturing exposure, gradually tilted the balance towards the discipline orientation that the former Harrison group advocated.

Research that was funded by the government, which was almost all of the fundamental and much of the advanced development work performed at the Princeton Laboratories, differed from commercially-oriented product support work in ways that affected the working lives of the researchers at the laboratories. More often than not the "scientists" enjoyed higher status and with it certain tangible advantages.

Theirs was often individual work for which credit could easily be assigned; it could be published quickly because it was destined for the public domain; it gained professional recognition for the individuals, and for the Laboratories as a scientific institution. Moreover those who performed the work received priority access to support services, and were rewarded with advancement in the Laboratories hierarchy. Even government work that was more closely linked to products tended to involve less tedious and exacting development activity than commercial product research support demanded. Government contracts had no need to emphasize the kind of cost reduction that was essential for products that would be produced in very high volumes.

Commercial research support became harder to conduct at the Laboratories because the conditions that surrounded it were comparatively so much less attractive than government work. Because commercial work had proprietary content either for RCA or for one of its licensees, it often had to remain secret for years. Those who performed such research gained professional recognition only when patents were issued, a matter that was determined principally by the dictates of RCA's patent and licensing activities. Such problem solving work was, in any case, somewhat hard to credit to individual researchers, for it typically involved a team effort. Naturally bright young recruits who wished to improve their professional standing in the outside world most often chose to sign on for projects that gave them the most professional exposure, and it became harder to find scientific specialists who were willing to engage in the cross-disciplinary team efforts that commercial product support tended to require.

The new recruits also made it necessary to adopt a new research management style. Old style laboratory directors had assigned specific research tasks, dictated the research methods that would be used, and supervised the work closely. Younger researchers expected to be able to pursue their chosen problems in their own way. To laboratory veterans who generally had some experience with the requirements of licensees and with manufacturing conditions, the new recruits' insistence on the primacy of theory over experience was maddening evidence of their lack of discipline and their impracticality.

AUTONOMY BECOMES ISOLATION

It was not hard in the first decade after the war for the research staff at the David Sarnoff Research Center to maintain reasonably good communications with the members of the advanced development groups in the RCA product divisions. Senior researchers in Princeton knew many of the senior divisional engineers personally, and most of the manufacturing sites were within driving distance of Princeton. Nevertheless a certain amount of rivalry did develop between Princeton and the divisions. The divisional technical staff were bound to envy the better conditions, pay and status of the people who worked at the "Country Club," as the Research Center came to be called.

In the mid 1950s the scientist generation began to move into leadership positions in the Laboratories. As this occurred at roughly the same time that RCA dispersed its manufacturing operations across the United States, a widening rift formed between the Research Center and the Divisions. Although the research management made continuing efforts to inform the divisions about the work its researchers were doing, the theoretical nature of the building block research seemed of small interest or importance to operating managers. Any product implications that might be embedded in the research reports were too uncertain and too far removed in time to be taken seriously. Further the Laboratories had gained a reputation in the divisions for having sponsored technologies that failed in the marketplace. Certain Laboratories-sponsored products during the 1940s and 1950s had either lost out to competitors in their development phase or had later been rejected by consumers. Two examples were the forty-five rpm record that met failure in the highly publicized contest with CBS's thirty-three rpm "long-playing" record, and the professional video tape recorder that had been beaten to the market by a little-known high technology company called Ampex in 1956.

The growing separation between the Laboratories and the divisions posed particular problems for the ability to transfer technologies between them. The Advanced Development departments which were set up in the divisions received little funding from their local managements and resented having to rely on the Corporate Research Division for their budgets. Increasingly they took the view, at RCA as at other companies known as the N.I.H. (Not Invited Here) syndrome, that the complications of adopting and adapting Princeton generated technologies into their own organizations were so great that it was easier to develop their own approaches from scratch.

Had the laboratories' source of research revenue remained inviolate and had the research needs of the company continued to be defined by the same people, the Research Center's isolation might have had limited repercussions. In fact it coincided with a strategic watershed for the company occurring in the late 1950s that called into question the whole role of research at RCA. First, in 1957–58 the corporation lost its major source of revenue from domestic package licensing when suits by Philco, Zenith and the Justice Department all ended in adverse judgments or negotiated settlements that went against RCA. The same year government funding for research and development came under fire in Congress which appeared to be changing in its attitudes toward research. While David Sarnoff as Chairman remained a staunch research supporter, the man he chose to succeed post-war chief executive, Frank Folsom, held different views. Elmer Engstrom, former head of the Laboratories and current head of Research and Engineering had long been regarded as Folsom's likely successor, but he was passed over for John Burns, a former consultant who was known to have reservations about the role of the corporate research center and its relationship to the rest of the company.

RCA's product most recently identified with the Laboratories added to its unpopularity in the rest of the company. Color television, Princeton's most important post-war research project intended for immediate commercial development, had by 1957 been on the market for three years, and had so far completely failed to gain acceptance. Several RCA divisions, including Home Instruments (receivers), Electronic Components (tubes), and the National Broadcasting Company, were keeping color television alive at great expense, while other manufacturers and broadcasters boycotted it. The financial drain associated with color had saddled RCA with profits that were well below the five percent return on sales, and was using cash that was needed to finance RCA's entry into the computer business. John Burns' response to these problems was to reorganize the company into different product sectors, to install new management at the Laboratories, and to reformulate the Research Center's mission. The Research Center was once again asked to emphasize new product research with special attention to computer support research.

The shift in the Laboratories' mission imposed from above had swift repercussions. Product oriented work placed heavy strains on existing manpower, strains that were all the more destructive in that the work was unattractive to a large segment of the research staff. Budgetary restraints, already introduced in anticipation of the loss of domestic licensing revenue, became more restrictive. Some key scientists did leave RCA at the time, others threatened to leave, and researchers who had had to interrupt on-going projects to divert to work on new products, were stretched to the limit. The change would turn out to be a temporary one, but it demonstrated to the Laboratories' management the risks of being spread too thin. Electronics as a field of application had expanded in too many directions for one organization to support all of them.

The new regimen was barely established when pressure on the Laboratories eased. New and renewed sources of outside funding for research gave Princeton a chance to revert to its former emphasis on a long-term research mission. Government R & D funding increased abruptly after the USSR successfully launched its Sputnik satellite. Unforeseen revenues also came from a sharp upturn in foreign licensing when RCA discovered the nascent Japanese electronics industry as a major new licensing partner. The prohibition against package licensing imposed by the courts had not extended to foreign licensees, and soon payments from Japanese companies began to make up for what RCA had lost from its former domestic licensees.

The belated success of color television gave a further boost to the Laboratories' fortunes. In 1960 the product passed the breakeven point after losing money for six years. In the same year John Burns left RCA with time remaining on his contract. Elmer Engstrom succeeded him as Chief Executive Officer, and together Engstrom and David Sarnoff publicly recommitted RCA to pursuing its traditional all-electronics strategy based on leadership in technology. Even though the company could no longer dominate its research field, it was recommitted to technological risk-taking and self-obsolescence.

STRATEGIC INITIATIVE PASSES TO PRINCETON

In terms of its role in the company, the RCA Laboratories had been offered a strategic reprieve, but the Laboratories' management realized that continued pursuit of electronic diversity could lead to unmanageable complexity for the company in business terms. It was time for the Research Center to adjust its own mission to suit changed circumstances. A key requirement of the mission was that it must ensure the stability so essential to a productive research environment. If Princeton did not want to be called upon to support too many of the short-term product and process development needs of the RCA manufacturing divisions, it had to find assured long-term sources of external funding,

research revenue for which it could claim sole credit. Government
funding was not only too unpredictable, but it was taking increasingly
burdensome amounts of administrative time and resources as more and
more reporting requirements came to be attached to funding. In any
case, the state of electronics technology itself seemed to dictate a new
research approach. Circuitry based on individual semiconductors was
giving way to integrated circuits that made "building block" research,
as such, less useful. The logic of the situation seemed to suggest that
the Research Center ought to put increasing emphasis on producing
revolutionary follow-on products on the scale of television.

The institutional needs of the Research Center itself pointed to certain kinds of product opportunities. Ideal opportunities were those that
matched the capabilities of the existing Laboratories. They had large
potential sales; and they were based on technology that was proprietary
to RCA. Only large business opportunities would be visible enough to
justify the attention of top management; and proprietary technology
would ensure the stream of revenues from overseas licensing.

The Laboratories ideal product was not necessarily likely to be the
kind of product that the divisions would prefer. To the divisions, major
new products generally meant unfamiliar markets, and new processes
entailed costly and risky start-up periods. To resolve this conflict the
Laboratories advocated a new venture concept by which a new business began its life outside the structure of any existing division. RCA's
recent disappointing experience with its infant semi-conductor business in the early 1960s seemed to support such an experiment in
management. Although the Corporate Research Center had produced
some important advances in semiconductor technology, the new business that was designed to allow RCA to capitalize on its research
achievements was not receiving the attention it needed. The fatal error
had been to locate it in the RCA electronics division where the overriding concern was to maintain the health of competing vacuum tube
products.

James Hillier, head of RCA's corporate research center in the early
1960s expressed the new role he had defined for the Laboratories in the
phrase "Laboratory as Entrepreneur."[27] By this he meant that within
its own structure the Laboratories should be able to carry out all phases
of the innovation process from idea generation to initial marketing.

It was easier to see that a new role was needed than to get the
research personnel to accept it. The new philosophy required both
administrative reforms and a shift in research priorities that struck at

the heart of the way projects were chosen and funds allocated. Directors of individual laboratories had to surrender some of their powers to decide what projects to back to a centralized planning function operating from Hillier's office. In 1965–66 the entire organization was restructured away from functional groupings by discipline into product-related groupings such as the Consumer Electronics Laboratory. Perhaps most important to the individual researcher, "sustaining research" so attractive to the theoretically inclined, had to be cut back; and the amount of effort that could be expended on exploratory projects before they had to demonstrate commercial feasibility was also reduced. Despite efforts to achieve the transition through persuasion and concensus, another period of upheaval ensued that undermined staff morale, threatened to drive away promising researchers, and reduced the productive output of the entire organization for a period.

The RCA Laboratories made only one serious attempt to act out the role of entrepreneur. It tried to launch a product of its own devising called Homefax soon after the internal reorganization took place, in 1966–67. Homefax was a home facsimile system, a forerunner to modern videotext systems. The laboratories had originated the product technology in its own Acoustical and Electronic Research Laboratory. It performed the market analysis and preliminary planning in a newly formed Business Evaluation unit. It even formed a venture group located in Princeton to launch the product. The company announced Homefax in 1967, and it was endorsed by David Sarnoff, still Chairman, and by Elmer Engstrom, C.E.O., as the kind of "systems" product that RCA had historically excelled at developing. But the Laboratories' full entrepreneurial mettle was never to be tested, for Homefax fell victim to another change in top management.

NEW LEADERSHIP, NEW ROLE FOR RESEARCH

In January, 1968 Robert Sarnoff succeeded Elmer Engstrom as RCA's Chief Executive Officer. As former president of NBC the younger Sarnoff's background, his interests and his sympathies were non-technical. He brought a team of professional managers from outside RCA who were oriented towards marketing and finance. The new head of marketing lacked enthusiasm for the complexities of the systems product Homefax represented. He saw little prospect that the product could be successful without the agonizing delays that had plagued color television in the previous decade, and he withdrew corporate support from the project. At a time when the financial markets looked favourably on companies who could introduce new technology based products, Robert Sarnoff and his management team were anxious to have new products to introduce, but they wanted them right away, ready for low-risk introduction on a compressed schedule. They were also indifferent as to whether they came from the research center or the divisions, or whether their technology was proprietary or not. Lacking experience with major product innovation, the senior management at RCA made demands on their research organization that went counter to the nature of revolutionary science-based innovation.

For the research organization at RCA the end of the Age of Big Science already portended a decline in support for long-term research. Now with the change in leadership the jobs of senior technical managers became increasingly political. They were working with technologies that could easily take ten years or more from first demonstration of feasibility to finished project. This was timing that non-technical managers neither appreciated nor wanted to tolerate, and even that pace depended on a stable research environment that top management was loathe to guarantee. As a consequence Laboratories management in particular had to spend more and more of its time in New York, selling its projects and then keeping them alive. It could often muster backing for its long-term projects from its powerful ally, the RCA Licensing Department which still accounted for a significant proportion of the company's profits and needed a steady stream of new proprietary technology to offer its licensees. But without the explicit commitment of top management to self-obsolescence in its product line, it was forced to compete on a year to year basis with the divisions for limited R & D funding. Further, it faced the dilemma that substantial incremental improvements in existing product and process technologies already in the divisions, when judged on a short-term basis, fortified the case against investment in revolutionary innovation. Even though the laboratories continued to provide significant amounts of research support for the divisions, the political activities of its management gave it the reputation of opposing and obstructing steady incremental development.

To review the argument, RCA was affected by the changing research climate around World War II in several ways: the increase in federal funding for research in industry and the competition it created for manpower, the perceived national need for industry to assume more of the weight of fundamental research, and the changing research philosophy that elevated theory to a higher status than application not only in academia but even in the eyes of many who were employed in industrial research establishments.

RCA's response to these climatic changes stemmed partly from management conviction, partly from opportunism and partly from sheer necessity. It created its central research laboratory, modelled on a university environment and staffed by a growing proportion of scientific specialists, to add to an already competent product and process research organization the ability to create revolutionary innovation. New business opportunities that arose of such efforts might necessitate some self-obsolescence, but it was assumed that the pace of change could be controlled in RCA's interest. There was also some naivete about the more negative aspects of performing government funded research side by side with commercially oriented work. The internal competition that went beyond creative tension, and the gradual isolation of the central research organization, were unforeseen consequences. Unintended too, was the tendency for those committed to revolutionary innovation to undercut when possible the ability of the divisions to institute continual changes in existing processes.

A key contextual factor, in addition to those of organizational structure and staffing, was obviously leadership. When David Sarnoff was still involved on a day to day basis he was able to promote a company commitment to long-term research, and to the institution that performed it, without letting it undermine advanced engineering support in and for the divisions. He understood instinctively the cultural and motivational differences between long-term and short-term R & D and he was able to force the necessary linkages between research and manufacturing divisions from above. Lacking his broad and eclectic background that combined sympathy for technical effort with a grasp of business essentials, his successors understood less well the philosophical and behavioral conflicts between research and other business pursuits. As a result they made unreasonable demands on research and forced it into a partisan role.

No one company can be regarded as wholly representative of any industrial phenomenon. RCA had clearly unique characteristics that in all probability exaggerated some of its peculiar contextual responses. Its existence for decades as a virtual research monopoly, for instance, was true for only a few companies, such as Alcoa in aluminum. Without question, at RCA the tradition of serving as primary research organization for an entire industry and the resulting ability to derive substantial revenues as a direct return for research output, gave the Laboratories an extraordinary attitude of independence towards its own product divisions. But the contextual elements that were important in RCA's response to climatic changes in research were also important in other companies and other industries. Without benefit of detailed or systematic research into other companies a few generalizations suggest themselves.

RESEARCH AS INDUSTRY

Post war changes in the industrial research climate were apt to have their full impact on the way research was conducted in companies only gradually. Their effect was both cumulative and interactive. But even by 1951 it was apparent that the infusion of large sums of government money, accompanied by the Big Science philosophical shift concerning the proper work of science, had turned industrial research into what many termed a $2 billion industry. If government funded fundamental research was regarded first primarily as a way to induce young research recruits to venture into industry, it soon came to be viewed as the possible first step in attracting substantial follow-on government business for manufacturing divisions. Research managers only slowly realized the full costs of performing, side by side, different kinds of research for different markets. These costs were a tendency to undermine established laboratory routine, the need for new management approaches, and heavy administrative burdens, and management complexity.

The new diversity in industrial research often led to organizational restructuring, as companies consolidated the supporting research from different businesses within related industries. Companies that had not formed centralized research laboratories before often found themselves obliged in any case to set up surrogate campuses if they wanted to keep their technical personnel. Unlike engineering problem solving, good research required a stable environment, the chance to work uninterrupted, and to be free of administrative concerns.

But as the RCA example demonstrates, once set up as independent entities and deliberately insulated from outside interference, central research organizations were apt to take on lives of their own with internal goals and institutional needs that sometimes differed from those of the parent corporations. It was in the nature of such institutions, particularly those staffed by a high proportion of scientists, to strive for autonomy, for control over their own research programs, and for the freedom to give their researchers the kinds of rewards they valued. These frequently were not the rewards that business traditionally offered its employees: money and promotion in the corporate hierarchy might be attractive, but they were secondary. Researchers cared more about professional standing, and this was determined by such matters as the institutional prestige of the organization that employed them, the chance to publish papers and to attend conferences, and the resources to support their work and to fund their increasingly costly research equipment. These differences in values were hard for most non-technical people in companies to comprehend, and they had a marked effect on the way their research organization related to the rest of the corporation.

After the early euphoric period when the benefits of Big Science were taken on faith in the United States, discontent about the low output and the general lack of control over research programs surfaced in management circles throughout the United States. Once separated from the operating locations, a corporate laboratory became a place in which change was undertaken only with the greatest reluctance. As long as those who managed the research organizations were familiar with the rest of the company the divergence in character was masked. But in the mid 1950s when the post-war generation took over industrial research leadership, open conflict often occurred between a laboratory and the rest of the company it served. When the research carried on at remote locations was accused of being practically irrelevant, research managers found it almost impossible to communicate the importance of their activities to people who did not already share the Big Science philosophy.

In the early 1960s a general shift occurred in the industrial context of research. All over the country the industrial reaction against the Big Science philosophy came to a head. The fresh infusion of government funding for Big Science that occurred after Sputnik caused the costs of industrial research to escalate dramatically, a combined consequence of increasingly complex equipment and disproportionately high scientific salaries. After creating inflationary pressures government funding reached a plateau. From 1967 industry had to pay more to sustain a level effort. At the same time a move towards professional management put pressure on operating divisions for steadily rising earnings and greater accountability. More and more heads of manufacturing divisions were evaluated as profit centers and were interested only in short-term product or process research support that would improve their immediate earnings picture.

In light of the adverse shifts in the industrial research context, managers of research, who had the task of matching the values, priorities and strengths of a discipline-oriented research staff to the needs of their corporations faced an almost insuperable task. If they were to see to completion the revolutionary product innovations that their decades of fundamental research had been supposed to engender, they had to find corporate avenues for business development outside the traditional ongoing businesses. It was the institutional necessity of research laboratories to promote long-range product innovation, and thus justify their existence in business terms, that led many research managers to argue for direct entrepreneurial role for research. Frequently this argument was based on breath-taking ignorance of the practical realities of beginning a new business. Nevertheless numerous executives accepted the view that innovation based on science should be treated as a system of honorable gambling where the costs of many losses could be covered by one hit on the scale of DuPont's nylon.[28]

The stakes in this kind of game were high, perhaps highest for the research organization itself, and the odds against success increased as the decade wore on. Although the consumer marketplace and the financial markets were receptive to new science-based products in the early 1960s, new businesses launched over the opposition of other operating divisions, or even in competition with them, had a high probability of failure. Perhaps the most bothersome legacy of this period for the corporate research center was the persistence of the gambling analogy. Long after the industrial context had come to favor a very different role for the research laboratory, and with it a different approach to research, research productivity in many corporations would still be evaluated subjectively according to the stark success or failure of a few visible projects, while credit for cumulative process improvement and product evolution would be accorded to divisional advanced development departments no matter where work was actually performed.

IMPLICATIONS FOR FURTHER STUDY

In the 1980s the United States finds itself in the middle of the first major shift in its research climate since the one that occurred in the Era of Big Science more than thirty years ago. We have heard arguments in support of increased government funding for R & D, but this time it has coincided with a different prevailing philosophy concerning the appropriate role for science in industry. Whether the institutional arrangements that translate funding into research programs will remain the same or are about to change, is still an open question.

The problem behind the decline in innovation in the 1970s, we are told, is that after decades of investment in science, U.S. industry has no shortage of fundamental scientific knowledge, but lacks the ability to apply it successfully. The accuracy of such a diagnosis is beyond the scope of this paper to determine, but the evidence is strong that such a belief has gained credibility with American industrial leaders. Under such circumstances the high value placed on applications will influence management behavior and possibly government policy in the U.S. for some time to come.

The question that has occupied us here is how individual companies and industries can be expected to react to climatic shifts of the sort described, and how they are likely to be affected in their ability to use the fruits of research. One way of distinguishing between likely responses is on the basis of the core technologies involved, particularly on their relative states of maturity. Chemically based industries were less radically affected by the extremes of Big Science than were companies in electronics, for instance. Further government funding for research in the 1950s emphasized electronics, and therefore Physics, to a much greater degree than chemically related technologies and disciplines. But companies within the same industry, based on the same core technology can have widely different reactions to broad climatic shifts, and these must surely be attributable to the kinds of contextual factors—leadership, organization, location and staffing—that we have identified in the study of RCA.

Further research is needed before the importance of context can be established for different kinds of companies. If we want to understand the contextual factors, as well as the government policies, that translate common environmental conditions into widely divergent research and innovation outcomes, nothing will substitute for detailed historical work in individual corporate research organizations. Such "applied history" is costly, painstaking and slow to yield results, but in the face of climatic changes for research that are likely to influence our ability to reap rewards from research for decades to come, the potential benefits of further historical works are likely to repay our investment.

NOTES

1. J. A. Schumpeter, *Capitalism, Socialism and Democracy* New York: 1950. Schumpeter conceived of the modern industrial research laboratory as an institutionalized substitute for the entrepreneur.

2. D. Noble, *America By Design: Science, Technology and the Rise of Corporate Capitalism* New York: 1977, 118.

3. John Servos in his article "The Industrial Relations of Science: Chemical Engineering at MIT, 1900–1939" ISIS 1980, 71 (No. 259) argues pursuasively however that even at MIT the close partnership that university scientists enjoyed with industry after World War I was a temporary phenomenon that soon gave way before the basic researchers' need for independence in problem choice and freedom of publication.

4. See for example Stuart W. Leslie, "Charles F. Kettering and the Copper-cooled Engine," *Technology and Culture,* 1981; D. C. Mowery, "The Emergence and Growth of Industrial Research in American Manufacturing, 1899–1945" unpublished PhD Dissertation, Stanford University, 1981; and L. S. Reich, "Research, Patents and the Struggle to Control Radio: A Study of Big Business and the Uses of Industrial Research," *Business History Review,* 1977 and "Industrial Research and the Pursuit of Corporate Security: The Early Years of G.E. Labs" *Business History Review,* 1980.

5. U.S. Senate Subcommittee on War Mobilization, *The Government's Wartime Research and Development, 1940–1944,* a study prepared by the Productivity and Technological Development Division of the Bureau of Labor Statistics. 79th Congress, 1st Session Subcommittee Report #5 to the Committee on Military Affairs of the U.S. Senate, March 1945.

6. National Research Council Bulletin #2 *Industrial Research Laboratories of the United States, 1921.*

7. C. Pursell, "Government and Technology in the Great Depression," *Technology and Culture* 20: (1979), 162–174.

8. D. Kevles, Ch. XVI, "Revolt Against Science," *The Physicists,* 236–281.

9. K. Compton, *New York Times,* February 24 and 25, 1934.

10. R. Kargon, p. 163 in *The Rise of Robert Millikan: Portrait of a Life in American Science,* Ithaca: 1982.

11. Science Advisory Board Report of Chairman, 1935, MIT Archives; and see *New York Times,* December 8, 1935.

12. E. U. Condon, "Recruitment and Selection of the Research Worker," p. 61 in *Scientific Research: Its Administration and Organization,* Washington, D.C. 1950.

13. A. Bright and J. Exter, "War, Radar and the Radio Industry," *Harvard Business Review,* Winter, 1947; W. R. Maclaurin, "The Organization of Research in the Radio Industry After the War," *Proceedings of the I.R.E.,* 1945.

14. U.S. Senate, Government's Wartime Research and Development, op cit., 5–11 and 20–22.

15. Ibid., 2–4.

16. D. Kevles, "The National Science Foundation and the Debate Over Postwar Research Policy," 1942–1945, *ISIS,* 1976.

17. U.S. Senate, "Government's Wartime Research and Development, p. 3. Also C. Greenewalt, speaking at ceremonies opening DuPont's Experimental Station in

1957 said, "We can no longer rely on basic science from foreign universities indulging our industrial genious, but without contributing our fair share to the world's store house of basic knowledge." Widener Library Pamphlet.

18. M. Trytten, Op. cit., *Scientific Research*, 70.

19. J. Morton, The effect that changing discipline mix had on leading industrial laboratories is illustrated by Bell Labs profile where by 1969 employees with advanced degrees numbered 20% of technical staff while, of the age group in its 30's, over 40% had Ph.D's. *Organising for Innovation*, New York: 1971.

20. T. Parsons, "Professional Training and the Role of Professors in American Society." Op. cit., *Scientific Research*, 150.

21. J. Conant, *Science and Common Sense* (1951), 305–309.

22. This account of the evolution of RCA's corporate research is based on my forthcoming book *The Business of Research: RCA and the Video Disc*. Note that until 1968 RCA was officially called the Radio Corporation of America (R.C.A.).

23. K. Kilbon, *Pioneering in Electronics* (unpublished history of research at R.C.A.), The David Sarnoff Library, located at the David Sarnoff Research Center in Princeton.

24. *The National Research Council, Bulletins #104 and 120 of Industrial Research Laboratories, United States.*

25. L. Hattery, "New Challenge in Administration," *Scientific Research*, Washington: 1950. "In a decade there has sprung full blown a research and development activity which calls for technically trained manpower beyond the supply, which renders equipment obsolescent more rapidly than it can be replaced, and which cries for administrative leadership not available."

26. See comments by H. Smyth and F. Harbison, both professors at Princeton University in "Industry and the Future of Basic Research," Princeton University Conference held in 1958. Both men warned of the dangers of industrial laboratories doing basic research to keep their personnel happy.

27. J. Hillier, "Venture Activities in the Large Corporation," *IEEE Transactions on Engineering Management*, June, 1968.

28. N. Fast, "Dupont Development Department Evolution From 1960–1976, ICCH study, Boston, MA: 1972.

CREATING A MONOPOLY:
PRODUCT INNOVATION IN
PETROCHEMICALS

Robert Stobaugh

A firm that is the initial commercializer of a petrochemical, by definition creates for itself a monopoly—strong or weak depending on the
competition of existing products. The rewards can be quite lucrative.
As of 1982, for example, one U.S. petrochemical manufacturer had
received a cumulative cash flow of $22 million from a single specialty
resin product, or some 50 times the company's cumulative cash out-
flow of $450,000 needed to develop and commercialize the product.
Indeed, one new product can make an important contribution to even a

Research on Technological Innovation, Management and Policy
Volume 2, pages 81–112
Copyright © 1985 by JAI Press Inc.
All rights of reproduction in any form reserved.
ISBN: 0-89232-426-0

giant's profits. Ciba Geigy earned over $1/2 billion in the United States alone from just one pesticide—atrazine. And DuPont's profits from nylon easily exceeded $1 billion and are believed by executives in the industry to have contributed as much as half of the firm's profits for years.[1]

Although it is clear that commercializing a new petrochemical product can be extremely lucrative, there is no study available that shows the average profitability received on funds spent to achieve this goal. The only systematic study of the subject indicates that profits of the firm that initially commercializes a new petrochemical are, on the average, good—but erratic. This study indicated that the average expected profit for the first three years of a petrochemical's commercial life was 2 1/2 times the funds spent to commercialize the product. But all these profits were accounted for by one-third of the products, for which the expected profits were eight times the funds spent on commercialization. The other 2/3s of the products in the aggregate showed a slight loss.[2] Unfortunately, there is no indication of the amount of funds spent to develop products that were not commercialized. And without some indication of this amount, the overall profitability of the funds spend in attempts to commercialize new petrochemicals cannot be known.

But what is known is that high risks accompany the potentially high rewards. The creation of a new product involves a journey that is long, difficult, and chancey. Prior to reading about this journey, some readers will want to know more about the petrochemical industry in order to judge the generality of the findings. Table 1 presents important characteristics of industry. The industry starts with a raw material obtained from crude oil or natural gas and produces end-products that are used mostly in synthetic materials. In fact, the industry is not one monolithic whole, but rather a series of many different subindustries each consisting of a slightly different set of producers producing an individual petrochemical, with some overlap between the markets. Products consist of both specialties and commodities, and range in age from new up to 60 years or more. The potential target of a successful innovation is large, for worldwide sales of the entire industry exceed $200 billion annually and sales of individual products can be several billion dollars annually; partially as a result, innovation has been relatively high compared with other industries.

ACTIVITIES INVOLVED IN PRODUCT COMMERCIALIZATION

In the petrochemical industry, the commercialization of a new product seldom is derived principally from a unique grasp of basic research. Rather, it results mostly from the large number of interrelated activities required to produce a product for a selected market.

True, basic research, which is the search for a fundamental understanding of natural phenomena, provides the foundation of knowledge that makes possible the commercial birth of a petrochemical. But scientists engaged in basic research generally prefer to offer their results as a free-good to their scientific peers, including those with other firms, universities, and government laboratories; and basic research is undertaken regardless of whether the knowledge sought has commercial value. A key attribute of the published results of science is that they can easily be understood by all state-of-the-art scientists worldwide; for if this were not the case, the results would not be accepted for publication.[3] Thus, the petrochemical industry spends less than 10 percent of its total research and development expenditures on basic research. But, even if a firm were to spend heavily for basic research, its needs, being quite diverse and unpredictable, would force it to rely on outside sources for most of its fundamental knowledge.[4]

Styrene monomer, a commercially important petrochemical from which polystyrene is made, can be used to illustrate the lack of commercial monopoly typically associated with basic research.

Styrene monomer was originally recovered by the distillation of a natural material called "storax," of which one source was dragon's blood, obtained from a Malayan rattan palm. The first reference to styrene monomer appeared in 1786 in a chemical dictionary.[5] Subsequently, articles appeared in the 1830's in French and German, including a famous one in 1839 by E. Simon, who named the product "styrol" (I do not know when the name was changed to styrene monomer). Simon also discovered that styrol polymerized to form polystyrene, later to become an important polymer. Subsequent research and publications followed, in English, as well as in French and German.

It was over half a century later, however, before serious efforts were made to commercialize styrene monomer. Naugatuck Chemical Com-

Table 1.   Important Characteristics of the Petrochemical Industry

Market Characteristics

Product differentiation	Specialties, which are products sold on the basis of performance, are differentiated; commodities, which are products sold on the basis of well-defined characteristics, are not[1]
Size of total market	Exceeds $100 billion in United States and several times that amount worldwide, with any single petrochemical accounting for a small portion of total[2]
Number of sellers and buyers	Sellers of a particular product typically few enough in number to constitute an oligopoly, but with more sellers of commodities than of specialties; typically more buyers than sellers[3]
Barriers to entry	Technology in early years of product's life, scale economies in later years[4]
Vertical integration	Some vertical integration between raw materials and petrochemicals and between some petrochemicals; less vertical integration between petrochemicals and products made from petrochemicals[5]
Conglomerateness	Each petrochemical made by a slightly different set of manufacturers than every other petrochemical, with some overlap in members of different sets; parent companies from many different industries—mostly oil and chemicals, but also steel, rubber, and others; most manufacturers make a number of different petrochemicals[6]

Supply

Raw materials	Petroleum products from crude oil and natural gas[7]
Technology	Some batch but mostly continuous process facilities that are highly automated, and capital intensive, with large economies of scale[8]
Work force	Proportion of engineers and scientists greater than the average in American manufacturing[9]
Value/weight	Cost of transporting a petrochemical typically small compared with its price[10]
Research and innovation	One of leading spenders on research and development and on output of innovations compared with other industries[11]

Table 1.   (Continued)

Efficiency and progress	High rate of productivity growth compared with other industries [12]
	Demand
Nature of end-product	A liquid or a solid that can be further processed in one or more steps into final products most often made from plastics, synthetic fibers, or synthetic rubbers [13]
Price elasticity	Low for specialties, and high for commodities—with specialty status most often occurring in early years of product's life and commodity status most often occurring in later years [14]
Substitutes	Final products made from petrochemicals compete in many separate markets, sometimes with one another and sometimes with products made from natural materials, such as wood, leather, silk, cotton, and rubber [15]
Rate of growth	High during early years of a product's life and low during later years [16]
Length of product's life	Can be 60 years or more [17]
Cyclical aspects	Correlated with business cycle—especially the commodities [18]

Notes:

1. See my *Petrochemical Manufacturing and Marketing Guide,* Vols. 1 and 2 (Houston: Gulf Publishing, 1966 and 1968); Frederick A. Lowenheim and Marguerite K. Moran, *Faith, Keyes, and Clark's Industrial Chemicals,* 4th ed. (New York: Wiley, 1975); Robert B. Stobaugh and Phillip L. Townsend, "Price Forecasting and Strategic Planning: The Case of Petrochemicals," *Journal of Marketing Research* (February 1975), pp. 19–29; and interviews with marketing executives.
2. George B. Hegeman, "Memorandum to the Petrochemical Energy Group," Arthur D. Little, Inc., Cambridge, Mass, August 1, 1981.
3. Reference 1, above.
4. Robert B. Stobaugh, *Petrochemical Manufacturing and Marketing Guide,* Vols. 1 and 2; Lowenheim and Moran, *op. cit.;* plus Robert B. Stobaugh, "Channels for Technology Transfer: The World Petrochemical Industry," in Robert Stobaugh and Louis T. Wells, Jr., *International Technology Flows* (Boston: Harvard Business School Press, 1984).
5. Robert B. Stobaugh, *Petrochemical Manufacturing and Marketing Guide,* Vols. 1 and 2; and Lowenheim and Moran, *op cit.*
6. *Ibid.*
7. *Ibid.*
8. *Ibid.*
9. U.S. Bureau of Labor Statistics, *Employment of Scientific and Technical Personnel in*

(continued)

Table 1.   (Continued)

Industry, 1962, Bull. No. 1418 (Washington: Government Printing Office, 1964) and
Employment and Earnings Statistics for the United States, 1909–64 (Washington: Government Printing Office, 1965).

10.   Robert B. Stobaugh, "The Product Life Cycle, U.S. Exports and International Investment," unpublished D.B.A. thesis, Harvard Business School, 1968, p. 295.

11.   For R & D expenditures, see Richard R. Nelson and Sidney G. Winter, "In Search of a Useful Theory of Innovation," *Research Policy* (1977), p. 39. For innovation, see A. Wade Blockman, Edward J. Seligman, and Gene C. Sogliero, "An Innovative Index Based on Factor Analysis," *Technological Forecasting and Social Change* (4, 1973), pp. 301–316.

12.   Productivity growth in the U.S. manufacturing sector (of 23 industries) was highest for chemicals in terms of percentage yearly increases in total factor productivity and output per worker. See Kendrick quoted in Nelson and Winter, "In Search of a Useful Theory of Innovation," p. 39, Tables 5.1 an 5.5.

13.   Reference 5, above.

14.   Interviews with marketing executives.

15.   Reference 5, above.

16.   Reference 5, above, plus Robert B. Stobaugh, "The Product Life Cycle, U.S. Exports, and International Investment," Chapters 1 and 2 and Appendix C.

17.   *Ibid.*

18.   U.S. Department of Commerce, *U.S. Industrial Outlook,* various editions (Washington: Government Printing Office, various dates).

pany, in the United States, tried around 1925, and failed. Badische Anilin and Soda-Fabrik (BASF, then part of I. G. Farben in Germany) succeeded in 1931, when it began making styrene monomer for use in the manufacture of polystyrene, which it commercialized at the same time. Thus, a century and a half after the original product had been discovered, BASF created a monopoly, but the monopoly was not based on an exclusive knowledge of basic research. Indeed, the innovator did not have a product patent.[6] Rather, the monopoly was based on "product innovation"—the set of activities that starts with the basic knowledge and ends with a new commercial product; that is, a new manufactured product of acceptable quality, available for sale by a company in a position to produce in quantities and at a price that would yield a profit.[7]

A classification of product innovation activities into stages, although somewhat arbitrary, is convenient for explaining the nature of the technological edge that a product monopolist in the petrochemical industry builds against its rivals. I use six stages of activities, some of which proceed concurrently.[8]

The first three, when added to basic research, are included as part of a firm's "research and development" activities:

Applied research, which involves searching for new scientific knowledge for a particular commercial use;

Preparation of basic specifications, which involves preparing a description of raw materials, products, and main manufacturing steps;

Design and operation of a pilot plant, which involves manufacturing a product in a small-scale plant in order to provide product for testing by potential customers and generate information to aid in the design of a larger plant. Although a pilot plant in the petrochemical industry is composed principally of vessels, pumps, piping, valves, and instruments, it fits the description of small-scale units used in other industries in that it is "flexible and highly reliant on manual labor and craft skills utilizing general-purpose equipment."[9]

These three activities together are commonly called "applied research and development." The "development" includes the technical development of a suitable manufacturing process and product. It also includes commercial development—especially market research, market development, and economic evaluations.

The next three stages involve:

Design and construction of commercial manufacturing facilities, which usually involves transferring responsibility from the research and development department to the engineering department and the operations department. This transfer of knowledge from a small-scale unit ("pilot plant") to a specially designed commercial manufacturing facility contrasts with a general description of the innovation process in which small-scale units *develop* into units that rely on automated, equipment-intensive, high-volume processes.[10]

Start-up of commercial manufacturing facilities, which includes training operators, writing operating procedures, and operating the plant until an acceptable level of product quality and output is reached.

Start-up of commercial marketing activities, which involves establishing a distribution system, training a sales force, and performing other marketing activities prior to the initial sale and delivery of the manufactured product.

A central feature of product innovation is the great uncertainty involved throughout all activities, but especially in the applied research and development stages. In fact, a prime goal of applied research and development is to create certainty out of uncertainty—order out of chaos. The high level of uncertainty is illustrated by one study of the innovation process for 17 petrochemicals: the three stages of applied research and development together accounted for three-fourth of the elapsed time, but only 43 percent of the total costs. The 43 percent was about equally divided among the three stages. In contrast to the long time required to complete the applied research and development, the design and construction of the manufacturing plant, although accounting for about 40 percent of the total cost, typically accounted for only one-sixth of the elapsed time. [11]

Product innovation in the petrochemical industry involves not only the development of a unique technology but also the matching of this technology to the needs of a market, a matching that requires heavy doses of the integration of many different kinds of specialists—scientists, engineers, operators, marketing researchers, and others. The integration process creates a need for frequent, effective, and swift communication among the specialists. [12]

To show the richness of real life not captured by aggregate data, I obtained records for a nine-year period for an attempted development of a new petrochemical by a U.S. company engaged in petrochemical R & D and manufacture. This company's disguised name is Alpha. Based on my experience in commercial chemical development while in industry, I judge that Alpha's journey depicts activities that are typical in attempting to commercialize a new petrochemical product. [13]

ALPHA'S JOURNEY

During the 9 years, Alpha's activities were confined to only two stages of the six stages of the innovation process; these two were *applied research* and the *preparation of basic specifications*. During this time, the firm obtained product from three different facilities: (1) a small facility in its laboratory that produced only ounces, called "laboratory scale" by the company; (2) a larger laboratory unit that produced 15 pounds in total, called "bench scale" by the company; and (3) an outside custom manufacturer that produced 500 pounds in total. At the end of year 9, Alpha was deciding whether to build a pilot plant with a capacity of 1 million pounds a year, a step that would be necessary to proceed towards commercialization. If the pilot plant were to be built and operated, then three stages would still remain—design and construction of manufacturing facilities, start-up of manufacturing facilities, and start-up of marketing activities. The alternative of building the pilot plant was to stop the project, and attempt to sell to another company whatever knowlege had been developed. (I do not know whether Alpha decided to continue the project.)

By the end of year 9, Alpha had spent some $2 million on the project, of which $1–1/2 million represented expenditures within that firm, principally salaries and related expenses. The other half-million dollars of costs had been incurred principally for services. The rate of expenditure had risen from $100,000 yearly at the beginning of the project up to $400,000 by year 9. The cost of a pilot plant would exceed $1 million and operating costs would be a least that much again. The cost of a commercial plant would be about $12 million, and the total costs of the product innovation perhaps $20 million. Many project development projects are much more costly than this one; for example, in an attempt to develop a new polymer, one firm in 1983 had spent $3.0 million over a 6-year period, reaching an annual spending rate of $1.2 million. It was expecting (hoping for) commercialization in 1988 by which time some $20 million would have been spent on research and development. A total of $70 million was projected to be spent through 1990, by which time profits were expected to be substantial.[14]

The project at Alpha started when a senior research scientist concluded from a survey of literature that a halide-based polymer had commercial potential for use in baked-on coatings. He speculated that such coatings could be used as fire-retardant wall covering. This was the beginning of year 1. A consultation between several members of the Research Department was followed quickly by a consultation between the Research Department and the Commercial Development Department. Like many firms operating in the petrochemical industry, Alpha had a separate department to manage the commercial development of new products. This department had the responsibility of integrating the activities of the different specialists involved in product innovation, including an integration of the long-run focus of research with the short-run focus of marketing.[15] Figure 1 shows those units of Alpha involved in the commercialization of petrochemicals.

As both the Research Department and the Commercial Development Department favored proceeding with the project, the General Manager of the Petrochemical Division gave approval to proceed with an evaluation. The ultimate hope, albeit quite hazy, was to commercialize a petrochemical product that could be used to make building materials that would have better fire-retardant characteristics than existing materials. The basic technology was to involve the reaction of halides and hydrocarbons, most likely through a number of process steps involving several reactions and attendant separations.

An analysis of the nine-year history of this project highlights the

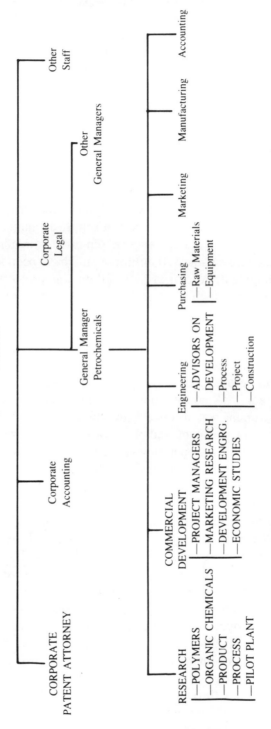

Figure 1. Organizational Units Involved in Commercialization of Petrochemicals, Alpha Company (Capital letters mean that a unit's principal activity is innovation, either product or process)

Source: Interviews with company officials.

importance of: (1) the great uncertainty encountered, and (2) the network of communications used.

Uncertainty

As the project wound its way through the thickets of uncertainty, almost all starting expectations, as well as many expectations adopted along the way, underwent major changes. The underlying causes of uncertainties can be classified into two categories: market and technology.[16] These two uncertainties, in turn, resulted in uncertainties about price, volume, and timing. Indeed, a range of raw materials was considered in studying how to make a range of possible compounds for a range of possible uses. The primary technical goals were the performance of a product in a given application and the efficacy of the process.

Two items are particularly important in determining the efficacy of a chemical process:

1. Conversion in the reaction, that is, the portion of the raw materials that actually undergo reaction. This is important because of the high costs involved in separating unreacted raw materials from the product.
2. Yield of the reaction, that is, the portion of the reacted raw materials that actually end up in usable product, rather than as a low-value by-product or waste.

Major variables in determining overall efficacy include type of catalyst, temperature and pressure of the reaction, and impurities in the raw materials.

The following examples illustrate the high degree of uncertainty encountered by Alpha.

Market: Initial target was baked-on coatings, but by end of year 9, non-baked coatings, hard elastomers, soft elastomers, rigid foams, flexible foams, and adhesives had been considered. Four different types of polymers had been evaluated by the end of year 9.

Technical: Twenty different processes had been evaluated involving dozens of different raw materials and 9

	different intermediate monomers from which polymers are made.
Price:	Estimates of possible selling prices ranged from 31 cents to $1.11 a pound.
Volume:	Plant capacities that were considered ranged from 500,000 to 50 million pounds annually.
Timing:	In year 3, a critical-path schedule was issued showing that a decision on commercialization could be made in March of year 4. But at the end of year 9, a commercialization decision appeared to be some years away.

As a result of the twists and turns of the project, it had required ten separate approvals by the General Manager of the Petrochemicals Division.

The Network of Communications

Communication links were within the firm; with customers, both actual and potential; and with other outside entities. At the center of the links was a Commercial Development Project Manager, who was appointed to coordinate the project.

This Project Manager had no authority over those in other Alpha departments and so had to rely on persuasion rather than commands. This lack of authority probably increased the number of links within the company since he did not have the luxury of setting up a direct chain of command through which his orders would flow. At any rate, he became the central figure in a complex communication network within the firm, as shown in Figure 2. In addition to the eight continual and five sporadic communication paths used by the Project Manager, there were a number of other communication links within the firm.

The parts of the communication network actually used during a given time period depended on the status of the project. Initially, the principal actors were the Project Manager and the Marketing Research Manager of the Commercial Development Department; Group Leader, Senior Scientists, and Analytical Laboratory personnel of the Research Department; and the accountants who provided a record of expenditures. From time to time—at least quarterly and often monthly—reports were given to, and discussed with, the General Manager.

As the project progressed, other groups entered the picture—sometimes on a continual and sometimes on a sporadic basis. By the end of

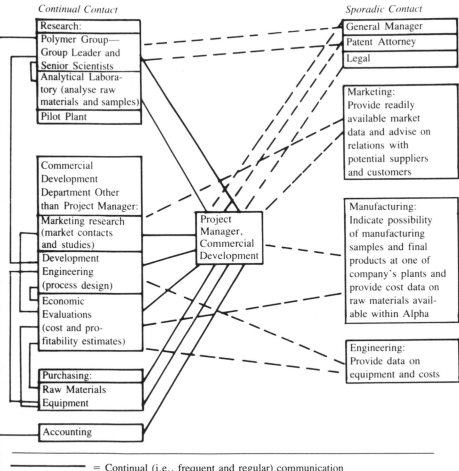

Figure 2. Principal Communication Paths within Alpha Company During Years 1 to 9 of Innovation of Petrochemical

year 9, all of the paths shown in Figure 2 had been used, many of them during the entire period. If the project were to proceed after year 9, additional communication paths would have had to be created as the project moved through additional stages. For example, the decision to proceed with a pilot plant would create a number of paths to and from the pilot plant section of the Research Department.

Alpha began to create a communications network outside the firm

very early in the project. During the first year, the Project Manager
made contact with an outside company—a combination of testing
laboratory and consulting firm—that evaluated the products and advised Alpha what the probable reactions of customers would be. During year 2, the Project Manager began to contact potential customers.
By the end of year 9, Alpha had had innumerable contacts with outside
firms.

Table 2 lists the principal contacts that Alpha's Commercial Development Department had outside the firm; but undoubtedly others
would be established if the project proceeded into year 10. Also, other
departments had numerous outside contacts. Altogether, the Commercial Development Department contacted over 150 different organizations, most of them numerous times. Most of the outside contacts,
whether counted in terms of numbers of organizations or numbers of

Table 2.   Contacts Outside of Alpha by Alpha's Commercial
Development Department

Type of Organization	Number	When	Purpose of Contact
Potential Customers	9 50	year 2 years 3–9	Determine interest in potential products
Competitors	more than 50	years 1–9	Obtain information, including samples, about potentially competitive products
Consultants	1 8	year 1 years 2–9	Obtain specialized services, principally market reports and economic evaluations
Custom manufacturers	2	year 7	Have a 500-pound sample of a potential product made
Testing Laboratories	4	years 3–9	Obtain tests of potential products
Engineering firms	1	years 3–9	Obtain estimates of plant costs
Others, including raw material suppliers	more than 25	years 1–9	Keep abreast of raw material prices and industry developmemts
Totals	more than 150		

Source: Alpha's records

separate contacts, were intended to reduce market uncertainty. Virtually all contacts with potential customers, competitors, and consultants had this as a goal. Contacts with the other four categories of organizations listed in Table 2 were intended to reduce technical uncertainty as well as market uncertainty.

Within the chemical industry, manufacturers typically send samples of proposed new products for the asking to any firms believed to be capable of developing commercial uses for them. So Alpha obtained samples of potentially competitive materials from three manufacturers. These three firms and other manufacturers of potentially competitive materials gave speeches at industry meetings, and Alpha attempted to gain as much information as possible through judicious questioning of competitors at such meetings. In order to facilitate these contacts and to keep abreast of raw material prices and industry developments, Alpha's commercial development personnel attended a yearly average of 12 professional meetings dealing with marketing research and commercial development.

One problem nagged at Alpha, and would continue to do so if the project went ahead: how to keep the communication links alive over many years given that individuals often change jobs. Alpha used frequent fact-to-face contacts as well as some written procedures to help keep internal links alive. But external links posed a greater difficulty. Here the commercial development manager had to maintain close and frequent contacts with the outside world to ensure that proper introductions were made when a newcomer replaced a person on the communication grid.

Alpha and the Theory of Innovation

Certain conclusions drawn from the Alpha case seem to break new ground. Other conclusions cover familiar territory, sometimes providing additional illumination, sometimes not.

The conventional model of innovation implies a simple linear progression from one step to another, such as:[17]

1. idea generation,
2. screening,
3. business analysis,
4. development,
5. testing,
6. commercialization.

In fact, because of *recycling* and *simultaneity,* this conventional model is quite inadequate in describing the real world.

There is a large amount of recycling from "later" stages back to "earlier" stages. Here is one example from Alpha: After the testing of baked-on coatings (step 5 of the conventional model), Alpha did additional screening (step 2) and business analysis (step 3). Then Alpha personnel generated a new idea (step 1): Try non-baked coatings. This new idea then moved forward through steps 2, 3, and 4 to arrive again at testing (step 5).

Furthermore, a number of steps often are conducted simultaneously. For example, Alpha personnel, when considering flexible foams, were screening (step 2), analyzing business opportunities (step 3), developing and improving the process (step 4), and testing potential applications (step 5) all at the same time.

Given that Alpha managers considered *seven* different target markets to be served by *four* different types of polymers made from *nine* different intermediate monomers, in turn made from *dozens* of different raw materials via *twenty* different processes, it is not surprising that the amount of recycling and simultaneity among the different steps was enormous.

The Alpha case does support some existing theory concerning the importance of environmental factors. First, it illustrates the importance of being able to rely on a diverse group of suppliers while a technology is in a fluid state. But Alpha shows not only the importance of the suppliers of goods (such as raw materials), which are emphasized in conventional theory, but also the importance of suppliers of services. The reader will recall that Alpha obtained services from nine consultants, four testing laboratories, one engineering firm, and two custom manufacturers. Second, Alpha illustrates the necessity of close contact with the market, an idea that is one of the linchpins of innovation theory.[18]

The Alpha case does, however, suggest that the often-accepted categorization of research as being either "demand-pull" or "technology-push" can be quite misleading, True, an Alpha research scientist did initiate the idea, but he also had in mind a possible market; and there was quite a bit of feedback from the market that generated new ideas. Hence, some observers might characterize this project as "technology-driven" with a strong check by marketing. In fact a better description is that it involved "backing and forthing" between demand and supply considerations.[19]

This "backing and forthing" highlights the amount of uncertainty inherent in innovation both at the level of the individual project as well as at the level of R&D project selection. Indeed, in Alpha's case, it is difficult to define whether there was just one project or whether there were many. And if there were many, how many were there: four, one for each different type of polymer: or seven, one for each market target; or twenty, one for each process; and so on. This "backing and forthing" combined with the uncertainty in defining the boundaries of a project means that managers, when taking actions on any one project or in selecting a set of projects cannot be maximizing in any serious sense of the word. Rather, they are engaged in search processes that are interacting; they aim primarily toward goals that are near rather than ultimate; and they use rules of thumb to guide their decisions.[20]

Finally, despite some indications that there has been a speed-up in the rate of innovation in industry in general, the Alpha case illustrates the long, arduous process that can be involved in the innovation of a new petrochemical.[21]

THE ROLE OF LARGE FIRMS

The process of product innovation is reminiscent of a journey of a gang of talented, unruly workers—optimistic dreams bouncing in their heads—with no one in clear command. Going, they know not how, to a destination they know not where. Further complicating the picture, the mix of workers is continually changing and many workers are part-time, devoting the remainder of their time to other projects. The endeavor obviously is risky.

The question naturally arises: what size of firm is best suited to complete such a journey? The answer, at least in the petrochemical industry, is quite clear: large firms.

A glance at the names of firms that were the first to begin commercial production of 31 principal classes of plastics reads like a list from "who's who" in the world chemical industry, and most of these firms were well-established major firms long before the introduction of the plastic in question.[22] True, in a few cases, usually in the early decades of the industry, some firms came into existence to produce a single product. In 1909, this was true of General Bakelite (later merged into Union Carbide), which introduced phenol-formaldehyde resins.[23] But relatively big firms introduced all nine petrochemicals for which I have extensive data on technology, investment licensing, and trade (here-inafter referred to as the Nine Products (see Table 3). Bayer, BASF, and Dynamit-Nobel were among the largest firms in the German chemical industry, and in the 1920s, they merged with others to form I. G. Farben. DuPont and Standard Oil of New Jersey (now Exxon) were the largest firms in their respective industries in the United States during the 1940s and Standard Oil of California was one of the largest in its industry.

Although there is no generally accepted theory about the relationship between the size of a firm and its level of innovation there has been a long-running argument on the subject. Some authors postulate that the stultifying effects of large size outweigh its advantages, whereas others maintain the opposite view.[24] The arguments about the advantage of size in the innovation process stem from Schumpeter and have been articulated well by Raymond Vermon, who particularily stresses the roles of risk and communication.[25] Certain of Vernon's explanations apply particularly to petrochemicals.

There is substantial risk associated with the large expenditures over the long period of time needed for a successful commercialization of a new petrochemical. The Alpha case is a good example. Included in this risk is the cost of specialized personnel and equipment.

The act of hiring highly skilled persons, such as those engaged in product innovation, can be considered as an investment decision because the firm initially loses money on such persons with the expectation of recouping it, and a profit, later. Most of the persons in the Alpha case, of course, were highly skilled. The sums involved are not trivial; the investment value of scientists, engineers, and general managers in the eight most skill-oriented U.S. industries is the equivalent of about 30 percent of the net worth of the firms. And the decision to hire skilled workers is risky, for the firm may have misestimated and not have sufficient use of their output, or they may not perform up to expectations, or they may leave or die.[26]

Even if small firms in the aggregate have the same record of research successes, and hence the same return as large firms, the large firms still have an advantage. Given the plausible assumption that a large firm will start more projects than a small firm, the proportion of successes for a single large firm is more closely predictable than the proportion for a single small firm. The greater predictability, of course, means lower risk for the same return, a condition preferred by investors.

Table 3. Selected Information about Original Commercialization of Nine Representative Petrochemicals

Year of Initial Commercialization	Name of Product	Innovator (Firm that Initially Commercialized Product)	Nation in which Innovator Headquartered	Nation in which Product Initially Commercialized
1907	Phenol	Bayer	Germany	Germany
1923	Methanol	BASF	Germany	Germany
1927	Vinyl chloride monomer	Dynamit-Nobel	Germany	Germany
1931	Styrene monomer	BASF	Germany	Germany
1933	Acrylonitrile	BASF	Germany	Germany
1942	Cyclohexane	DuPont	U.S.	U.S.
1944	Isoprene	Standard Oil of N.J.	U.S.	U.S.
1945	Ortho-xylene	Standard Oil of Calif.	U.S.	U.S.
1949	Para-xylene	Standard Oil of Calif.	U.S.	U.S.

Source: Trade journals and correspondence with companies
I consider them to be representative of basic and intermediate petrochemicals, which are used primarily in the manufacture of plastics, fibers, and rubbers. The products are made from raw materials that can be transported internationally; hence, the manufacture of the product is not tied to the availability of a natural resource.

In addition to the above arguments about risk made by Vernon two
other arguments about risk apply to petrochemicals (and probably certain other industries).

First, the large firm is favored by the desire of prospective users of a
new product to reduce their risk. This desire to avoid risk causes
prospective users to favor large firms as a supplier of a new product.
They argue that DuPont is a more reliable supplier than a small, and
perhaps unknown, chemical company. Even if this were not the case,
the user's purchasing agent is less likely to have problems with other
members of his or her firm if DuPont, rather than the small, unknown
company, fails to deliver a quality product.[27]

Second, the large scale economies in the commercial manufacture of
petrochemicals favor large firms. These scale economics set a lower
limit on the size of a plant—the "minimum economic size." The cost
of a plant of minimum economic size represents a relatively larger
investment, and hence a larger risk, for small companies than for large
ones.[28] This risk is made even greater because of the specialized nature
of many plants producing petrochemicals commercially; a plant often
cannot be converted easily to the production of other chemicals if the
new product proves to be unprofitable. Of course, a small firm could
develop the product and then sell the technology to a large firm for
commercialization. But this has rarely been done in the petrochemical
industry, not surprisingly, because the market for knowledge is notably imperfect.[29]

In terms of efficient communications, there are advantages for a
firm to have within itself the specialists and specialized equipment
needed for product innovation:[30]

- a high degree of secrecy is involved and secrets are easier to keep
  if outsiders are not involved;
- face-to-face communication is the preponderant means for trans-
  mitting information and ideas, and this typically takes place easier within a firm than across firms;
- most persons involved in product innovation—scientists being
  an exception and they are a small minority—are more receptive
  to signals received from within their firms than from outside; and
- the specialists and specialized equipment will be more readily
  available if located within the firm rather than being purchased as
  needed.

True, firms do rely on some outside assistance for R&D activities. But insofar as the mix of capabilities and equipment needed for effective research is indivisible, as is often the case, a large R&D program is more likely than a small one to achieve a higher degree of specialization within the firm.

Alpha's case provides insight into the type of activities that can be contracted to outsiders. In general, Alpha used outsiders principally for very specialized information or services for which it was not economical for Alpha to generate within the firm. For example, Alpha obtained from consulting firms studies of existing markets and available technology, with which Alpha's personnel were not familiar. Although this information was specialized from the viewpoint of Alpha, it was information that the consulting firms either had available or could generate readily from their knowledge and files. In addition to purchasing information from outsiders Alpha had products tested by outside laboratories that specialized in testing polymers and building materials. And Alpha had limited quantities of products made by custom manufacturers that happened to have equipment of the appropriate size.

Note that these tasks for which Alpha contracted with outsiders had two characteristics, in addition to being specialized. First, they were not part of indivisible activities, that is, they were sufficiently well-defined so that a price for the service could readily be determined. Second, they were not the heart of the innovation. Although the innovation would represent the result of a large and varied set of activities, the heart of the innovation was the laboratory work in which proper manufacturing conditions and product specifications were sought. These activities remained within Alpha.

In addition to advantages in risk and communications, the large firm has two other advantages over the small firm. One is in the use of knowledge generated in research. The large firm, with its wider field of activities, is more likely than a small firm to be able to find a commercial application for a research discovery. True, the small firm could sell the knowledge, but, again, the small firm would encounter the problem of the imperfect market. The second advantage is that the greater number of new product introductions possible for large firms compared with smaller firms results in lower introduction costs because of the greater experience obtained.[31]

The facts cited that large firms were the main innovators of 31

principal classes of plastics and the Nine Products seems to be about
the strongest evidence available that confirms the advantage of large
firms over small firms in commercializing petrochemicals. But this
conclusion also is supported by a historical description of the commercialization of organophosphorus insecticides, which are petrochemical
compounds made from phosphorus and organic chemicals.[32] On the
other hand, one can draw very few firm conclusions from the enor-
mously large economics literature on the relationship between the size
of a company and its level of innovation. This literature, rather than
focusing on product innovation, deals with innovation in general—that
is, process as well as product innovation. Furthermore, the relatively
little of it that deals explicitly with petrochemicals contains contradictory conclusions, albeit the bulk of the weight tends to support the
"large-firm advantage" for petrochemicals.[33]

The relationship between firm size and product innovation in other
industries is even less certain than in petrochemicals. The principal
product innovators were large firms in video cassette recorders, both
large and small firms in semiconductors, and small firms in processed
foods.[34] Econometric studies, which are on innovation in general
rather than on just new products, also show ambiguous results.[35]

Rather than arguing that large has an advantage over small, or vice
versa, in innovation in general, a more fruitful approach is to ask: what
situations within an industry and what conditions across industries
favor innovation by large firms over small firms, and vice versa?[36]
And how can large firms organize for innovation in order to take
advantage of bigness while minimizing its disadvantages?

THE IMPORTANCE OF LARGE MARKETS

Just as a large size gives an advantage to firms in commercializing new
petrochemicals, so does a large size make a nation more likely to be
the site of the innovation. One might suppose that the commercialization of new products might occur in a country with low costs for R&D
personnel and facilities—Colombia, for example. But this has certainly not been true for petrochemicals, which, by and large, have been
commercialized in large market countries. Indeed, 46 of the 53 most
important plastics, synthetic fibers, and synthetic rubbers were first
manufactured commercially in either the United States or Germany.
The United Kingdom, France, and Italy accounted for six of the re-

maining seven.[37] And all Nine Products were first manufactured commercially in pre-war Germany or in the United States (see Table 2), the two countries with the two largest domestic markets at the time the products were introduced. The first five were commercialized in Germany, reflecting that country's early lead in organic chemicals and plastics. The last four reflected the U.S. surge in petrochemicals during and after World War II.

And in every case—for the 53 plastics and the Nine Products—the original commercializer was headquartered in the nation in which its initial commercial facility was located. A body of research explains why innovation of new products in general—not just petrochemicals—tends to occur in nations with a large market for any given product.[38] This conclusion holds for the two broad sets of activities involved in product innovation—applied research and development and start-up of commercial facilities.

Applied Research and Development

Large-market countries tend to be the location of the applied research-and-development activities leading to product innovation because considerable communication occurs between the firm's web of specialists and its customers—the link between the marketing specialists and the customers is obvious. But, in addition, there is contact between the customers and the personnel in the R&D facilities producing samples. In the case of petrochemicals, the customers are manufacturers of petrochemical products or end-products made from petrochemicals. There are more such customers in large-market countries with large populations and high incomes than in small-market countries; because, at least through 1985, such customers tended to locate their manufacturing facilities near the end-markets rather than in countries that might have lower-cost materials or labor.

Within the firm, the great uncertainty involved in applied research and development creates a need for managerial decisions at key points; thus, the web of specialists are likely to be located in the same metropolitan area as general management. Many companies in the U.S. chemical industry, for example, immediately after World War I, chose New Jersey as a location for their R&D facilities so that these facilities would be near their national headquarters in New York.[39] (An exception to this tendency is that pilot-plant facilities are sometimes located near large chemical plants or refineries in order to have on-site a source of a variety of raw materials. Exxon's pilot plants at its Baton Rouge refinery are an example.)

The result is that a trio of centers in a communication network are located in one country: customers, applied R&D facilities, and company headquarters.

The Location of the First Commercial Plant

Since the firm initially commercializing a product is a monopolist, its customers are not very sensitive to the price of the product. Some authors have concluded that because of this price insensitivity, the firm tends to locate the first commercial plant in the country with a market for the product because it has no particular incentive to look outside where production costs might be lower.[40] But even if a firm were to do a thorough search of other countries, it still might reach the conclusion that locating the initial plant in the consuming country is the best solution once communication needs and risk reduction are considered.

The operators of the plant must communicate frequently with each of the three centers that are involved in the applied research-and-development phase: R&D personnel, general management, and customers. Furthermore, it is efficient to have the R&D personnel readily available to form the nucleus of the commercial plant's work force, at least during the early periods of operation. The transfer of technology between an R&D group and operations is especially costly when the transfer is across national boundaries, because of the greater distance—in language, customs, and miles—that are typically involved.[41]

The need for good communications between the plant personnel and customers is especially severe in the case of solid petrochemicals, that is, polymers, elastomers, and fibers, which must meet a host of specifications. The list of characteristics that are important can be quite long, and each can be the subject of discussions between plant personnel and customers. Take polymers, for example: impact resistance, durability, stability, weathering, set time, adhesion, flexibility, gloss, appearance, hardness, light stability, abrasion resistance, acid resistance, drying time, initial color, color after exposure, and water resistance. Product improvement efforts and discussions with customers continue throughout the life cycle of such products, although more of them occur early in the life of a product than later.

But some petrochemicals, typically the basic and intermediate products from which the petrochemical end-products are made, are not solids. Instead, they are either gases or liquids. And gases and liquids, compared with solids, typically have fewer characteristics that are important. Purity is the primary criterion; minor impurities in a mono-mer, for example, can have a detrimental effect on the quality of any polymer made from that monomer. The gaseous and liquid petrochem-icals reach a "standardized " quality earlier in their lives than do the solids. But, even in the case of gases and liquids, there is still much negotiation—particularly on the question of impurities—between the manufacturer and the customers about product quality for some time after the initial commercial plant has been built.[42]

Such negotiations result in high marketing costs for new chemical products in general; according to one estimate, these costs average about 23 percent of the sales price during the introductory phase and decline to about 2 percent of sales price as a product reaches matu-rity.[43] There is every reason to believe that petrochemicals follow some such path. Locating the initial commercial plant in a country with a large market minimizes these early marketing costs.

Also important is the availability within the country of a large sup-ply of outside specialists likely to know their home market and the availability of materials suitable for manufacturing products for that market.

The key communication links in the foregoing description of the innovation process are shown in Figure 3. This is a highly simplified chart. Each link appearing on the chart represents many individual communication paths. For example, the single link connecting the network of product-innovation specialists to customers can represent dozens of actual communication paths, for a number of different spe-cialists can be in contact with any given customer and there can be many customers.

The desire to avoid risk, as well as to ensure good communications, reinforces the tendency for the first commercial plant to be located in the same country as the other three communication centers—applied R&D, customers, and general management. First, the large economies of scale in petrochemical manufacture set a lower limit on the size of a plant. Thus, locating the plant in a country with a large market instead of in a country with small one avoids the risk of depending on the export market for a large percentage of the plant's output; the pos-sibility that the importing nation could erect trade barriers makes an export market riskier than the domestic market. Second, there is quite

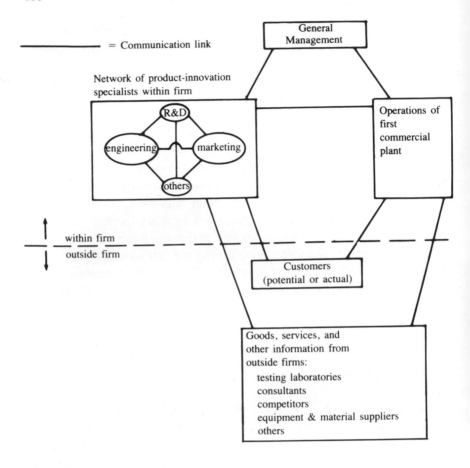

Figure 3. Simplified Conceptual Scheme of Communication Web Required for Innovation of New Petrochemical Products, Within and Outside the Innovating Firm

a bit of evidence suggesting that executives—even those of multinational enterprises—prefer to manufacture in their home country because of the inherent risks in operating abroad.[44]

Thus, a concatanation of forces, especially involving good communications and risk reduction, gives large-market countries a decided advantage over small-market countries as the location of the first commercial plant manufacturing a new petrochemical. And the facilities are likely to be owned by a large firm headquartered within that country.

It is the large nations and large firms, therefore, that by and large reap the monopoly profits derived from commercializing new petrochemicals.

* * *

Success at creating a monopoly by commercializing a new petrochemical rests not on a unique grasp of basic research, but rather on the completion of a long, complex, and chancey journey. This journey is not a simple maximizing exercise by a firm; instead, it is an interactive search process involving large amounts of "backing and forthing" between technology and market considerations in an attempt to reduce uncertainty along these two key dimensions. Two networks of communications—one within and one outside the firm—are crucial to success. Because of the risks encountered on the journey and the many types of resources needed, big firms have an advantage over small firms. And big nations have an advantage over little nations because of the great importance of close communication with the market.

ACKNOWLEDGMENTS

I benefited from the advice of Max Hall, Raymond Vernon, and a number of colleagues in the Production and Operations Management Area at Harvard Business School, especially William Abernathy, Kim Clark, Therese Flaherty, and Richard Rosenbloom.

NOTES AND REFERENCES

1. Interviews with company executives.

2. Edwin Mansfield and John Rapoport, "The Costs of Industrial Product Innovations," *Management Science* (August, 1975), 1380–1386. There is, however, some evidence that certain types of chemical product innovations have a low private rate of return. See Edwin Mansfield, et al., "Social and Private Rates of Return from Industrial Innovations," *Quarterly Journal of Economics* (May 1977); and *The Production and Application of New Industrial Technology*, New York: W. W. Norton, (1977).

3. For a more thorough discussion, see Therese Flaherty, "Information as a National Economic Resource," *Conference Proceedings, The International Congress on Applied Systems Research and Cybernetics*, 1981.

4. National Science Foundation, *Trends to 1982 in Industrial Support of Basic Research,* Special Report NSF 83-302, 1983. Also, see *Chemical and Engineering News* (September 18, 1972), 8: and Keith Pavitt, *The Conditions for Success in Technological Innovation* (Paris: OECD, 1971), pp. 79–103.

For further discussion of the use of outside source of knowledge see Richard S. Rosenbloom, "Product Innovation in a Scientific Age," *New Ideas for Successful Marketing* (Chicago, Illinois: Proceedings of the 1966 World Congress, American Marketing Association, 1966), Chap. 23; and Raymond Vernon, "Location of Economic Activity," in John H. Dunning (ed). *Economic Analysis and the Multinational Enterprise* London: George Allen & Unwin, 1974. For a study of an earlier time period, which shows that DuPont relied on inventions made outside the firm for the basic knowledge in 10 of 13 commercially significant products, see Willard F. Mueller, "The Origins of the Basic Inventions Underlying DuPont's Major Product and Process Innovations, 1920–1950," *The Rate and Direction of Inventive Activity* Princeton, New Jersey: Princeton University Press, 1962, p. 343.

5. This history is from A. J. Warner, "Introduction," and R. H. Boundy and Sylvia M. Stoesser, "History," in Ray H. Boundy, Raymond F. Boyer, and Sylvia M. Stoesser (eds.), *Styrene: Its Polymers, Copolymers, and Derivatives* New York: Reinhold, 1952, pp. 1–12; and Ernst von Meyer, *History of Physical Chemistry* London: Macmillan, 1906, p. 542.

6. It is my impression that patents are less important in providing monopoly protection for commercializers of relatively simple molecules, such as styrene monomer, than for commercializers of complicated petrochemicals, such as polymers, elastomers, and fibers. Patents are also important in certain non-petrochemical sectors of the chemical industry in which the molecules are very complex; for evidence on pharmaceuticals, see Eric von Hippel, "Appropriability of innovation benefit as a predictor of the source of innovation," *Research Policy.* (11/1982), pp. 95–115.

7. Definition adapted from Edwin Mansfield, et al. *Research and Innovation in the Modern Corporation* (New York: W. W. Norton, 1971), Chapter 6, who refers to both new products and new processes as "innovations", also see Edwin Mansfield and John Rapoport, "The Costs of Industrial Product Innovations," in *Management Sciences* (August 1975), pp. 1380–1381; and Donald G. Marquis, "The Anatomy of Successful Innovations," *Innovation* 7 (1969), 28. But some other authors refer to a commercial development as an "innovation" only if it results in a new product: see *Plastics: Gaps in Technology* (Paris: OECD, 1969), p. 17.

8. These six were developed in Mansfield, et al., *op cit.*

9. William Abernathy and James Utterback, "Patterns of Industrial Innovation," *Technology Review* (June/July 1978), 6.

10. *Ibid.* True, in the case of some petrochemical products that are specialties and have been used in small volumes—certain fluorohydrocarbons, as an example—it is possible for the initial commercial facility to be an expanded semi-commercial facility, which, in turn, was an expanded pilot plant. Source: Example obtained from interview with company executive.

11. The author refers to them as unnamed chemicals; from their general description, I concluded that all 17 were petrochemicals, reference 8, and Mansfield and Rapoport, *op. cit.,* p. 1382. A 1982 study reports that expenditures in commercialization (the last 3 steps in my description) for American industry in general have been

declining, from one-half of new product expenditures in 1968 to one-fourth in 1981, whereas the portion of expenditures in exploration, screening, and business analysis (perhaps equivalent to the "applied research" step in my description) more than doubled (10% to 21%) in the same period: see Booz-Allen & Hamilton, Inc., *New Products Management for the 1980s* (New York, N.Y., 1982), pp. 6, 12, 15. I have no information as to whether the pattern in the petrochemical industry has changed.

12. Rosenbloom, *op. cit.* and Vernon, in Dunning (ed.), *op. cit.* Close integration between R&D, marketing and economic evaluators increases the probability of commercialization; see Edwin Mansfield and Samuel Wagner, "Organizational and Strategic Factors Associated with Probabilities of Success in Industrial R&D," *Journal of Business* (April 1975) 179–198. Also, see Paul Lawrence and Jay Lorsch, *Organization and Environment* (Boston, Division of Research, Harvard Graduate School of Business Administration, 1967).

13. I performed the integrating role, described in Lawrence & Lorsch in reference 12 as a commercial development manager for Monsanto in the early 1960s; but the Alpha case is not Monsanto.

14. Alpha estimates from company records and interviews with an Alpha engineer. Estimates on the polymer commercialization from records of the other company.

15. For a description of this function, see Lawrence and Lorsch, *op. cit;* and Ruedi and Lawrence, "Organizations in Two Cultures," in P. Lawrence and J. Lorsch, *Studies in Organization* (Homewood, Ill: Richard D. Irwin, 1970), which describes the difference in approaches to this role in the German and U.S. cultures.

16. This two-category classification is discussed in Abernathy and Utterback, *op. cit.,* p. 7. The role of uncertainty is stressed in Nelson and Winter, "In Search of Useful Theory of Innovation," *Research Policy* (6/1977), pp. 47, 71.

17. These steps are adapted from Booz-Allen & Hamilton, Inc., *op. cit.* pp. 3, 11. Other authors use somewhat different steps, for example, scientific discovery, invention, development, innovation (commercialization) and diffusion (broad application), see William J. Abernathy & Phillip L. Townsend "Technology, Productivity and Process Change," *Technological Forecasting and Social Change* (7/1975) p. 380–81, who state that the linear model "is certainly valid in the sense that events do occur in the suggested sequence," although they do caution that it "is more misleading than useful." For still another exposition of a linear model, see S. Meyers and D. Marquis, *Successful Industrial Innovations,* National Science Foundation, 1969; Report No. 69-17.

18. The importance of suppliers of goods is emphasized in Max Hall, (ed.), *Made in New York* (Cambridge, Mass: Harvard, 1959) and is implicit in the "uncoordinated" and the "fluid" stages described in Abernathy and Townsend, p. 390, and Abernathy and Utterback, p. 2. Certain case studies of innovation have highlighted the importance of the market, but typically the extent of contact with the market is not documented; for an example, see Leonard S. Reich, "Industrial Research and the Pursuit of Corporate Security; The Early Years of Bell Labs," *The Business History Review* (Winter 1980), pp. 504–29. The importance of the market is also emphasized in *Made in New York* and a long line of research flowing from that work, see Raymond Vernon, "International Investment and International Trade in the Product Cycle," *Quarterly Journal of Economics* (May 1966), pp. 190–207 and *Sovereignty at Bay*

(Basic Books, 1971), and Louis T. Wells, Jr. (ed.), *The Product Life Cycle and International Trade* (Boston: Harvard Business School Division of Research, 1972). For a separate line of research that reaches similar conclusions about the importance of the market, see N. R. Baker, et al., "The Effects of Perceived Needs and Means on the Generation of Ideas for Industrial R&D Projects," *IEEE Trans. Eng. Management* (December 1967), pp. 156–63; Meyers and Marquis, *op. cit.;* and Abernathy and Townsend, *op. cit.*—although the latter study contains an interesting twist by pointing out that one firm's *product* innovation often represents a *process* innovation for the target market.

19.   For a discussion of "demand pull" and "technology push" see C. Freeman, *The Economics of Industrial Innovation* (Harmondsworth: Penguin, 1974), pp. 108–9; Keith Pavitt, *op. cit.,* pp. 52–55; and Richard S. Rosenbloom, *op. cit.* The "backing and forthing" pattern is hypothesized in Richard R. Nelson and Sidney G. Winter, *op. cit.*

20.   This argument is found in Nelson and Winter, *op. cit.* pp. 51–53.

21.   There is conflicting opinion about the speed-up in the rate of innovation. See F. E. Burke, "Logic and Variety in Innovation Processes," in Ed. M. Goldsmith, *Technological Innovation and the Economy* (New York: John Wiley and Sons, 1970); and Edwin Mansfield, *The Economics of Technological Change* (New York: W. W. Norton, 1968), p. 102.

22.   *Plastics: Gaps in Technology, op. cit.,* p. 97. Major product innovations are obviously more appropriate than patents as this measure. The use of patents is a questionable measure for innovation in general. Union Carbide, for example, had a policy for years of not filing for patents but instead attempted to keep its operations highly secretive—to the extent of calling its units by code names and calibrating its instruments so that the plant operators would know neither the product they were making nor the operating conditions under which it was made.

23.   Such results are consistent with the argument of Dennis C. Mueller and John E. Tilton, "Research and Development Costs as a Barrier to Entry," in *Canadian Journal of Economics* (November 1969), pp. 570–579, except that stages of development apply to a whole industry rather than a product.

24.   See a brief description and references in Mueller and Tilton, *op. cit.;* Edwin Mansfield, "Size of Firm, Market Structure, and Innovation," *Journal of Political Economy* (December 1963), p. 556; and F. M. Scherer, *Industrial Market Structure and Economic Performance,* second edition (Chicago: Rand-McNally, 1980), pp. 413–418; and A. Cooper, "R and D is More Efficient in Small Companies," *Harvard Business Review* (June 1969).

25.   Vernon, of course, relies heavily on the research of others and has an excellent bibliography that I will not repeat here. See his "Organization as a Scale Factor in the Growth of Firms," in Jesse Markam and Gus Papanek, *Industrial Organization and Economic Development* (Boston, MA: Houghton Mifflin Co., 1970), and Raymond Vernon, *Storm over the Multinationals* (Cambridge, MA: Harvard University Press, 1977), Chapter 3.

26.   Vernon in Markham and Papnek, *op. cit.*

27.   B. David Halpern, "Growth Problems of a Small R&D Oriented Specialty Company," in E. Balgley, P. S. Gilchrist, P. B. Slawter, *The Small Chemical Enterprise and Forces Shaping the Future of the Chemical Industry* (American Chemical

Society, 1973), p. 43. I have seen this factor operate first hand in my consulting work with product innovators. Also see E. K. Bolton, *Development of Nylon,''* *Chemtech,* July 1976, p. 463; and Rowland T. Moriarty, *Industrial Buying Behaviour* (Lexington, MA.: Lexington Books, 1983).

28.   Robert B. Stobaugh and Phillip L. Townsend "Price Forecasting and Strategic Planning: The Case of Petrochemicals," *Journal of Marketing Research* (February 1975), pp. 19–29. Also see W. W. Alberts and S. H. Archer, "Some Evidence on the Effect of Company Size on the Cost of Equity Capital," *Journal of Financial and Quantitative Analysis,* (Vol. 8, March, 1973).

29.   See my "Channels For Technology Transfer: The World Petrochemical Industry," in Robert Stobaugh and Louis T. Wells, Jr. *International Technology Flows* (Boston: Harvard Business School Press, 1984); and P. J. Buckley and M. Casson, *The Future of the Multinational Enterprise* (New York: Holmes and Meier, 1976), chapter 2.

30.   Vernon in Dunning, *op. cit.,* Vernon in Markham and Papanek, *op. cit.;* Thomas J. Allen, *Managing The Flow of Technology* (Cambridge, MA: M.I.T. Press 1977); Richard S. Rosenbloom and F. W. Wolek, *Technology, Information and Organization* (Boston: Division of Research Harvard Business School, 1967).

31.   In a survey of 13,000 new product introductions between 1976 and 1981 in 700 large U.S. manufacturers, a 71 percent experience curve was found; i.e., with each doubling of the number of new products introduced by a company, the cost of each introduction declined by 29 percent; see Booz-Allen & Hamilton, Inc., *op, cit.,* p. 18. Also see Jean-Paul Sallenave, *Experience Analysis for Industrial Planning,* (Lexington, MA: Lexington Books, 1976).

32.   *Interaction of Science and Technology in The Innovative Process: Some Case Studies,* Battelle Columbus Laboratories, Report for NSF under Contract NSF-C-667, March 19, 1973, Chapter 9.

33.   See Edwin Mansfield, *Industrial Research and Technological Innovation* (New York: W. W. Norton, 1968), p. 42, for evidence that in the chemical industry there are scale economies in R&D activities (increases in R&D expenditures are associated with more than proportional increases in inventive output, when the effect of firm size is held constant by statistical methods) but when R&D expenditures are held constant, size of firm is associated with decreases in inventive output. In contrast, another study found no statistically significant relationship between firm size and R&D inputs in those industrial chemical firms that perform R&D; see William S. Comanor, "Market Structure, Product Differentiation, and Industrial Research," *Quarterly Journal of Economics* (November, 1967), 641.

Comanor studied 21 different industries and included only those firms that performed R&D. His results suggest that there are more industries of the 21 he studied in which all firms undertake greater proportionate levels of R&D than large firms than there are industries in which the reverse holds, i.e., large firms undertake proportionately more R&D (p. 641). For opposite results, see an earlier study of 17 industries using somewhat broader industry classes than Comanor, in D. Hamberg, "Size of Firm, Ologopoly and Research: The Evidence," *Canadian Journal of Economics and Political Science* (February 1964), pp. 74–75. Another author reported ambiguous conclusions about chemicals but tended to support Comanor's results about industry in general, see F. M. Scherer, "Firm Size, Market Structure, Opportunity and the Output of Patented Inventions," *American Economic Review* (December 1965), 1108–11, 1121.

34.   Sony, Matsushita, and Philips, the three first companies to commercialize video cassette recorders were all very large, see Richard S. Rosenbloom and William J. Abernathy, "The Climate for Innovation in Industry," *Research Policy II,* (1982), pp. 215, 217. In semiconductors IBM, GE, RCA, and Philco were large firms, and Texas Instruments and Fairchild were small firms, see John E. Tilton, *International Diffusion of Technology: The Case of Semiconductors,* (Washington, D.C.: Brookings Institution, 1971). For the food study, see Robert D. Buzzell and Robert E. Nourse, *Product Innovation in Food Processing: 1954–1964,* (Boston: Division of Research, Harvard Business School, 1967).

35.   Scherer, *op. cit.,* Chapter 15.

36.   This idea is mentioned in Therese Flaherty, "Issues in Assessing the Contribution of Research and Development in Productivity Growth: Comment," *Studies in Industry Economics,* No. 108, (Stanford University, October 1979). For a framework for viewing systematic differences in innovation among industries, see Michael E. Porter, "The Technological Dimension of Competitive Strategy" in Richard S. Rosenbloom (ed.), *Research on Technological Innovation, Management and Policy,* Vol. 1 (JAI Press, 1983), pp. 1–33.

37.   *Plastics: Gaps in Technology, op. cit.,* p. 97.

38.   Vernon, *Sovereignty at Bay, op. cit.,* Ch. 3 and G. C. Hufbauer, *Synthetic Materials and the Theory of International Trade* (Cambridge: Harvard University Press, 1966).

39.   Raymond Vernon, *Metropolis 1985,* (Cambridge, MA.: Harvard University Press, 1960) and Robert Lichtenberg, *One-Tenth of a Nation,* (Cambridge, MA: Harvard University Press, 1960).

40.   See Vernon in reference 38.

41.   David Teece, "Technology Transfer by Multinational Firms: The Resource Cost of Transferring Technological Knowhow," *The Economic Journal* (June 1977).

42.   H. M. Corley (ed.), *Successful Commercial Chemical Development* (New York: John Wiley & Sons, 1954), pp. 142 and 347.

43.   Roger Williams, Jr., "Why Cost Estimates Go Astray," *Chemical Engineering Progress* (April 1964), p. 18.

44.   Robert B. Stobaugh, *Nine Investments Abroad and Their Impact at Home* (Boston: Division of Research, Graduate School of Business Administration, 1976).

AMPEX CORPORATION AND VIDEO INNOVATION

Richard S. Rosenbloom and Karen J. Freeze

This is a story of lost opportunity—the story of how Ampex Corporation tried, and failed, to establish itself among the leaders in making and selling mass-produced video recorders. Founded during World War II on the peninsula south of San Francisco—just north of the area that would later gain fame as "Silicon Valley"—Ampex gained prominence and prosperity in the 1950s through engineering achievements that established it as the leader in the emerging technology of magnetic recording. In 1956 Ampex astonished the broadcast industry by demonstrating the first practical videotape recorder (VTR). In 1962, some 75% of all television recorders in use worldwide carried the Ampex name. Throughout the 1960s, Ampex invested in R & D to maintain its lead in this technology. It pioneered in development of the new helical formats and in innovating in the new markets they made

Research on Technological Innovation, Management and Policy
Volume 2, pages 113–185
Copyright © The President and Fellows of Harvard College.
ISBN: 0-89232-426-0

possible. Ampex, in the 1960s, was well positioned to claim its share of the growth of those new markets. Yet in the 1980s, when the VTR business worldwide grew to exceed $6 billion annually, Ampex's share of that revenue was less than 3%.

There is no necessary relationship between technological leadership and commercial success. But Ampex had more than technological excellence to bring to emerging business opportunities in the video recording field. It had a strong established commercial position in the field—effective marketing organizations operating worldwide and a brand recognized for quality by professional users everywhere. Add the financial strength provided by an initially strong balance sheet and the result should have been a formula for sure success. It wasn't, and it is instructive for us to inquire into the reasons why.

To some extent the answer must lie in the strengths of the competitors who ultimately seized the opportunities that might have been Ampex's. As we have argued in another place, the several Japanese firms that came to share more than 90% of the world market were persistent in executing sound, long-term strategies aimed at achieving a strong position in the market for mass-produced VTRs when it emerged.[1] The American firm best positioned to compete against these rivals was Ampex. Even under favorable circumstances, Ampex may not have been able to long sustain its early dominance of the industry. But even a 20% share would have yielded revenues exceeding a billion dollars annually in the early 1980s—several times Ampex's actual revenues from all products.

In order for the Japanese firms to win the competitive battle for the home video market as overwhelmingly as they have, it was also necessary for Ampex somehow to lose. By attempting to elucidate why and how this happened we can, perhaps, gain some insight into larger questions concerning the role of technological innovation in the growth of industries and the fortunes of companies. Post mortem analysis of business affairs should, of course, be interpreted with caution. The main actors have little incentive to help clarify the record of events; as John F. Kennedy observed, "victory has a hundred fathers and defeat is an orphan." Conclusions on such matters cannot be beyond dispute, as cause and effect blend together in the complex world of organizational life. Most important, the nub of the matter often lies in a necessarily speculative judgement about "what might have been." Mindful of these difficulties, we turn to an account of how Ampex got into the business of making and selling video recorders and how it tried, and failed, to follow that business into larger markets.

A PATTERN OF INNOVATION

The source of Ampex's growth and success in the 1950s was a series of stunning technological innovations. The force behind those innovations, and the intelligence that guided them, came from Alexander M. Poniatoff, the company's founder. Within the short span of five years, Ampex invented and commercialized professional quality sound and data recorders, and initiated the project that would result in the first practical television recorder. The commercial consequences were dramatic; sales grew at an annual rate of 70% (compounded) for a dozen years. Poniatoff's small engineering shop of 1948 was transformed into a substantial industrial enterprise with nearly $70 million in sales and 4,000 employees in 1959.

A consistent pattern characterized these innovations, exhibiting a style that would endure in the company's consciousness, if not always in its behavior. Ampex started from mastery of a robust core technology: magnetic recording. It applied that technology to develop a diversified line of recording equipment; each product was specialized to meet the needs of a particular customer group, such as broadcasters or aerospace engineers, for recording particular sorts of information, such as music or telemetry from rocket tests. Behind the diversity was a common thrust: Ampex made machines that were the first of their kind and that provided outstanding performance. Ampex specifications usually set the standard of performance in each field, and the company positioned itself at the high end of each market, where it became the dominant supplier.

The company's commitment to excellence was signalled at its inception, when Alexander M. Poniatoff formed the name from his own initials plus "EX" for excellence. Poniatoff had come to the United States in his mid-thirties, worked for 17 years as an engineer in large firms, and founded his own firm in San Carlos, California in 1944, when he was 52 years old. The Ampex Electric and Manufacturing Company made high performance motors and generators for naval radar systems. After the war, Poniatoff searched for a civilian product that would keep his small firm in business.[2]

He found that product serendipitously and thereby launched his firm as a pioneer in the novel technology of magnetic tape recording. In so doing he demonstrated a keen appreciation of the potency of the new technology and a sound instinct for a rewarding commercial application, qualities that would be exhibited recurrently in the next decade.

In November 1946 he was intrigued by a demonstration of a German Magnetophon, a magnetic tape recorder that Signal Corps officer Jack Mullin had recovered in 1945 from a German radio station. Poniatoff hired Harold Lindsay, the young engineer who had enthusiastically told him about the Magnetophon, to head up a project to develop a similar machine.[3] Lindsay and one of Ampex's electrical engineers, Myron Stolaroff, set to work on a playback head for Mullin's machine—without access to its electronics. Their remarkable success within a few months quickly led to the first American-made professional tape recorder, which Ampex dubbed the "Model 200." Upon seeing and hearing the prototype, Bing Crosby—who had tired of live broadcasts—immediately ordered 20 machines, which he subsequently sold to the ABC radio network. The first unit was delivered to ABC on April 24, 1948 and placed in service for national broadcasts the next day. Within a year Ampex had introduced an improved design, Model 300, which became the standard of the radio broadcasting industry.[4]

Following the success of the first product, Poniatoff took two important steps, one financial and organizational, the other technical and commercial. A growing enterprise needed capital and greater breadth in management resources. He chose to acquire both in a transaction with a local partnership of private investors, the Ayala Associates, in October 1948. Their equity investment gave them control of the company.[5] Moreover, under an unusual contract—in force to mid-1958—Ayala provided management for the firm. One partner, George Long, became an active executive, later succeeding Poniatoff as Chief Executive. Two others, Joseph and Henry McMicking, served during several decades as members of the Board of Directors.

The company's next step marked out a commercial strategy that would endure: Ampex would not be a specialist in broadcast equipment but would exploit its technical capabilities in diverse markets. In 1951 Ampex introduced the first in a line of instrumentation equipment. These recorders, selling for $3,000 to $30,000 each, were capable of precision recording of data on multiple tracks at frequencies as high as 100,000 cycles per second, five times the level required for audio information. They provided the first practical methods of acquiring, storing, and processing data in a wide range of scientific, military, and industrial applications.

In the next five years both the audio and instrumentation businesses prospered, producing roughly equal shares of revenue that grew, in total, from $1 to $10 million annually by 1956. Audio applications were extended to motion pictures (1953) and later to consumer products. Data recorders were adapted to airborne use (1954) and then as memory devices on large digital computers (1956). While this was going on, a small group of engineers were working on an innovation that would be the most dramatic and the most important in the company's history—a recorder for television broadcasters. The video project most clearly illustrates the pattern of innovation that came to be considered the norm at Ampex.

Ampex's Video Project

By 1951 the television boom was well underway in the United States; more than 11 million households already had receivers and the number grew by nearly a half-million every month. Standards for color broadcasting were being debated hotly, and the inception of coast-to-coast network transmission was on the horizon. The only available means for time-delay broadcasts for the West Coast was kinescope, a photographic technique of adequate quality but excessive complexity and cost.

In September 1951 David Sarnoff challenged the engineers at RCA to develop a videorecorder, which he termed a "videograph," within the next five years. Although crude recordings of television had been made on discs a generation earlier, most engineers assumed that magnetic recording technology would provide the best path toward realization of Sarnoff's concept.

A magnetic recording system includes sophisticated mechanical and electronic elements, but it is the magnetic components—the recording medium (coated tape, disk, or card) and the transducer (the "head" that imprints and reads the magnetic record on the medium)—that lie at the heart of the technology. Unless they perform well, the system will not. Also important are the scanner and associated transport mechanism, through which the head scans the medium to "read" or "write", and the electronic circuitry, which both processes the signals to and from the head and controls the mechanical actions of scanner and transport.

The first critical design choice in developing a videorecorder pertained to the type of scanner; there are three basic "families" of approaches:

1. "longitudinal scanners" in which the medium is scanned along its length by moving the tape past a fixed head at a velocity equal to the desired writing speed;
2. "transverse scanners" in which the medium is scanned across its width by heads rotating (at the writing speed) in a plane perpendicular to the (much slower) motion of the tape;
3. "helical scanners" in which the tape is scanned in a pattern slanted across its width by heads rotating in a plane set at an acute angle to the motion of the tape. (See Figure 1 for illustrations.)

The problems of obtaining adequate heads and tapes applied equally to all three approaches; the differences arose in the associated problems of mechanical design and electronic circuitry. The common requirements for all three scanner approaches were: (1) frequency response—to attain a high enough "writing speed" to record the high frequency video signal with then available heads and tapes; (2) time-base stability—to attain a very high degree of stability in the scanning motion (a sudden variation of one part in a million would disrupt the video image); and (3) standardization—to maintain the geometry of the scan over time and across machines so that a recorded signal could reliably be recovered at a much later date or by a different piece of equipment. Each of these depended critically on the mechanical design of the scanner and its electronic control circuits.

The engineers at RCA chose to develop a recorder using fixed heads. Given the limits of head technology at that time, they had to employ very high writing speeds—i.e. the tape moved past the heads at 30 feet per second (25 times the speed of the best audio machines). The resulting problems of control made compatible recording and playback extremely difficult. Moreover, only a few minutes could be stored on a huge reel, posing problems of space and expense still overwhelming in 1953. Nevertheless, RCA persisted and was expected to succeed in time.[6]

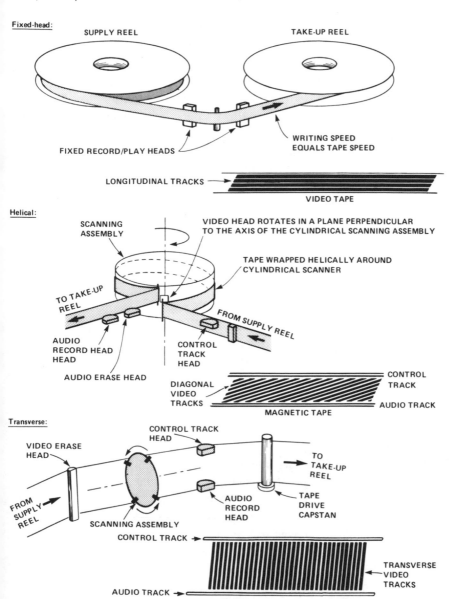

Figure 1. Videorecorder Scanners

Early in Poniatoff's career, General Electric had made him project engineer for a new type of circuit breaker because, he recalled their telling him, "experienced engineers already know that it cannot be done. You are not smart enough yet to know it is impossible." That spirit informed the work of the engineers who undertook to develop a video recorder at Ampex, a firm that had just reached the million dollar level, in competition with the giant Radio Corporation of America.[7]

In 1951 Ampex engineer Myron Stolaroff had been inspired by a demonstration of an experimental method of video recording developed by Marvin Camras at the Armour Research Institute (now the Illinois Institute of Technology). This new configuration embodied a rotating head across which tape passed at a reasonably slow speed. Crude as it was, it pointed to new directions for research.[8] Poniatoff, Stolaroff and another top technical aide, Walter Selsted, decided to hire a young broadcasting engineer, Charles P. Ginsburg, to head the development of such a machine. Ginsburg's varied academic background included engineering and physics; as for practical experience, he had spent several years working with the technical problems of broadcasting for stations in the Bay area.[9]

Ginsburg began in January of 1952 to explore the Armour idea, which employed three heads on the plane surface of a drum, rotating in an arcuate pattern across a two-inch tape. After a few months he was joined by college student Ray Dolby;[10] together they produced an "almost recognizable picture" in October 1952, and, with a machine using four heads, a much improved but still problematic picture in March 1953. As Ginsburg recalls, "the complete concept of an improved system burst upon Ray's consciousness one Sunday morning in October 1952 . . . and he came to my house to discuss it with me immediately." Dolby, choosing to work full time on the project, lost his student status and was drafted into the Army. Ginsburg was assigned to a new supervisor, with whom he came into conflict. In the summer of 1953 the project was shelved at Ginsburg's request.

The project was formally reinstated in July 1954. Ginsburg and Charles E. Anderson, who had joined the company in April 1954, had been tinkering with the design, with and without management's blessing. By mid-summer they had decided on two fundamental changes in the system. First, the heads, now attached to the convex surface of the drum, would no longer write in an arcuate sweep, but in straight lines perpendicular to the movement of the tape, which was cupped against the rotating drum by a guide mechanism. The second change was to be an automatic gain control system (a feature of the electronic circuitry).

Later that fall Fred Pfost and Alex Maxey joined Ginsburg and Anderson. Pfost had come to Ampex two years earlier, right after graduation from the University of California, and was attracted to the challenge of developing a suitable head for recording high frequency signals. Maxey, then 28 years of age, had worked for the Army in audio recording and for the Atomic Energy Commission as a technician. Raised on a ranch, Maxey left school after the ninth grade, began his career as a production worker, and discovered through experience the pleasures of invention. When he heard about Ampex's project through the "electronics underground" of the Bay area, he managed to see Ginsburg and convince him that he could help with that effort.

Early in 1955 Dolby returned from his stint in the military. The five-man team that would eat and sleep video recording for the next 14 months was now complete. In December 1954 Anderson, convinced of the superiority of an FM encoding system, persuaded Ginsburg to let him work on that instead of the automatic gain control. In three weeks he had a picture; by February 1955 he had buried the old AM format. Dolby subsequently streamlined Anderson's key FM innovation. On March 2 the team impressed Ampex's board of directors with a machine that had the basic elements of the future VR-1000 or "Quad," so named after its four heads.[11] Management now gave the project full support, aiming for a public demonstration in a year. The only formal study of market potential followed this decision. Its pessimistic conclusion that demand over the first three years would amount to only 30 units (if they were priced as low as $30,000 each) seems to have had little effect on management's determination to proceed.[12]

The months that followed constitute a story akin to that of many such projects: intense work, frazzled nerves, sleepless nights followed by technological breakthroughs. It was at this time, Ginsburg recalls, that Poniatoff gave the team "what we needed most: isolation from management."[13] One crucial challenge was to control the distance between the tape and the spinning heads; the problem was tricky because the distance varied as the heads wore down, yet had to be kept constant for accurate playback and, eventually, for interchangeability between machines. The distance was determined by the guide that held the tape against the drum; Maxey figured out how to control the position of the guide with a variable vacuum between it and the tape.[14]

Another recalcitrant problem, the delicate magnetic heads, fell to Fred Pfost. He completely redesigned the heads to increase their response, make them mechanically more durable, and enhance their manufacturability.[15]

The day of reckoning for the five engineers was in February 1956, when a prototype was demonstrated before a group of 30 Ampex employees. The demonstration astounded the audience. Management decided to commit the company to a public demonstration that spring. During the next two months several visitors, sworn to secrecy, viewed the machine; meanwhile, the team continued their efforts with renewed zeal to improve the picture quality and reliability—until the last moment preceding the first public demonstration on April 14, 1956.

This well-chronicled event was a spectacular success.[16] The place was Chicago, at a meeting of some 200 CBS affiliates preceding the opening of the convention of the National Association of Radio and Television Broadcasters. As TV cameras covered CBS chief engineer Bill Lodge's speech, the Ampex Mark IV—as the experimental model was named—recorded it. Within a few seconds the speech was played back, stunning the audience. Suddenly the room erupted in wild applause and cheers, as people pressed forward to look at the machine and to place their orders—pushing them into Philip Gundy's hands as he wrote receipts for the $45,000 machines "on the backs of envelopes."[17] Lore has it that before rushing to inform their associates of the exciting breakthrough, many in the audience called their stockbrokers.[18] Some 100 machines were ordered during the next four days of the convention.

AMPEX 1956–1960: COPING WITH SUCCESS

Euphoria reigned at Ampex in the glow of publicity immediately following the sensational performance in Chicago. Engineers and managers revelled in their new image as the David (800 employees and $10 million in sales) who dramatically wounded the RCA giant. But more than celebration was in store for the happy group of engineers when they returned home from the NAB Show in 1956. They had to develop two further technical improvements: interchangeability of tapes, since broadcasters had to be able to play back on any machine a tape recorded on any other; and color—while color broadcasting was still in its infancy, the demand for color recording capability was already emerging.

Others in the company faced the overriding challenge of initiating volume manufacture of the VR-1000, as the machine was now called, in a very short period of time. The television networks were anxious to put the new machines to work in place of kinescope for rebroadcast of national programs on the west coast. The broadcasters expected cost savings sufficient to repay the cost of the equipment within a year.[19] Ampex was also anxious to start installing machines, since that would create a *de facto* technical standard that other vendors would be forced to emulate.

As an interim step, Ampex offered the networks a few custom-manufactured prototypes for shipment in 1956 at a premium price of $75,000. Eleven of these went to the three U.S. television networks. CBS broadcast the first magnetically recorded time-delayed network program, "Douglas Edwards and the News," on November 30, 1956. NBC followed in January 1957, ABC in April when Daylight Saving Time began. (see Figure 2.)

Deliveries of the first production models of the VR-1000 were promised for the end of 1957, an ambitious goal. Unlike the prototype, these units would offer full interchangeability of tapes. The key to that lay in achieving consistency in volume production of the rotating video heads. With a usable life of only 100 hours, the heads had to be produced in volume; yet transferring the art of the engineer's touch to an assembly-line standard was no easy task.

Ampex's success in solving unexpected difficulties clinched its role as leader in the industry and hurled it into spectacular growth. By mid-1959 Ampex had sold more than 375 VTRs, and by 1961 nearly 900, representing some three-fourths of all videotape recorders in use worldwide. Nearly 300 were outside the United States.

Solving the problem of interchangeability was an integral part of the urgent challenge of making the VR-1000 truly manufacturable. Achieving color capability, if less pressing, was equally important in the long run. After exploring one promising approach, Ampex turned to the leader in color television, RCA. The two companies reached an agreement whereby they freely exchanged technical (though not manufacturing) know-how for three months, later extended to four months. At the end of 1957 Ampex granted RCA a license to use the patented inventions embodied in the VR-1000 in exchange for licenses under

The Ampex Videotape Recorder, installed in a leading West Coast television network facility.

Figure 2. Above: The VR-1000, 1956. (Annual Report, 1957)

124

Right: The heart of the VR-1000, its head assembly. (Annual Report, 1959)

RCA's color patents. Ampex produced its first color converter, the 1010 Kit, in 1958, and a time-base corrector for improved color in 1961. RCA began to deliver quadruplex recorders in 1959 and by 1961 had 25% of the market.[20]

The Japan Connection: Sony

With the achievement of color recording, the next logical technical improvement for the VR-1000 was to replace the awkward vacuum tube circuitry with solid state designs using transistors, then an emerging technology. A solid-state VTR would be more compact and more reliable. No U.S. company had as much experience with transistors suitable for this purpose as had the Japanese companies, who were the world leaders in transistor radios and were on the verge of developing fully transistorized television receivers. Ampex also wanted to gain access to the Japanese market, where television broadcasting had begun only three years earlier.[21]

Ampex Japan, Ltd., was established in early 1959 as a sales branch. Ampex was hoping to turn it into a manufacturing division like its counterpart in England, Ampex Electronics, Ltd., but Japan's Ministry of International Trade and Industry (MITI) would not permit such a venture without a Japanese partner. In 1959 Ginsburg, Maxey, and Selsted joined Philip Gundy, then head of the new Ampex International Division, on a journey to Japan to make preliminary inquiries. They saw several experimental video recorders at Sony Corporation—

including a crude replica of their own VR-1000. During the next several months they discussed with Sony an exchange of technology— Ampex's VTR know-how and patented inventions for Sony's expertise in transistors. In February, the general managers of several Ampex divisions visited Sony; in its Annual Report published in May 1960, Ampex stated that Ampex Japan was "currently engaged in the preliminary work necessary to produce the videotape television recorder early next year for Japanese and other Far East markets." In July 1960 Ampex and Sony exchanged memoranda on a contract for mutual technical assistance, and Ampex engineers went to Japan. In August, Sony's vice president, Akio Morita, visited Ampex to discuss the transistorization of the VR-1000.

Although Sony subsequently did assist in the development of Ampex's first transistorized VTR, the VR-1100, further cooperation was hindered by differences over Sony's claims—on the basis of a letter from the head of Ampex International—to the right to use Ampex's patented inventions without royalty payments. In mid-1961, Ampex's Annual Report would say only that "previous plans to assemble specialized products at Ampex Japan, Ltd., have been terminated. Ampex Japan, Ltd., as a corporate entity, is in a stand-by status, exploring further opportunities in this part of the world." Not until 1964 did Ampex announce further progress in Japan—an agreement with Toshiba to form a joint venture company for manufacturing in Japan.

The Managerial Challenge of Success

The skills required to manage Ampex changed dramatically in the mid-1950s. Within a few years' time, the small, privately-owned, engineering-oriented firm was transformed into a large public company operating in diverse markets. Earlier, innovative engineering had been sufficient for the company's success, when Ampex had taken the lead in applying magnetic recording to meet needs in several markets. Once established in those markets, the company had to be able to operate ongoing businesses and to fend off the challenges of competitors.

Growth was explosive. Sales tripled in two years and more than doubled in the next two, rising from $10 million in the year ending April, 1956 to nearly $70 million in Fiscal 1960. In the same period, net earnings grew thirteenfold, from $300,000 in 1956 to $4 million in 1960. The number of employees grew to 4,500 by mid-1960.

The executive responsible for leading the company through this eventful period was George Long. A former banker, Long joined the company as Treasurer in 1950 when Ayala Associates made their investment. He became executive vice president and general manager in 1953 and president and chief executive officer in 1956, after the video project was well underway.[22]

The company's growth in the late 1950s came both from expansion of its original businesses and entry into new markets. Sales of professional audio products, providing a majority of the company's revenues until mid-1956, continued to grow, but at a modest pace. Demand for instrumentation products boomed, quadrupling between 1956 and 1958, as defense and space programs generated robust demand for ground station and airborne instrumentation. Adaptation of Ampex designs to the special needs of digital computers for business use created a new line of business for the company.

The video recorder innovation, a source of much of the company's growth after 1957, followed the pattern by which Ampex had entered its other major markets. But Long also led the company into new markets where it lacked any distinctive technological edge. In 1957, Ampex entered the consumer audio equipment business and began the manufacture of magnetic recording tape.[23] The consumer audio market was very different from Ampex's other businesses, as more than 50 domestic manufacturers and a host of others competed for relatively few consumer dollars. In 1959, moving even farther from Ampex's established skills, Long created United Stereo Tapes to produce and distribute recordings for home listening.

The computer products business was reorganized and strengthened in 1960, when Ampex acquired Telemeter Magnetics, Inc., a pioneer in the production of ferrite core memories. The Ampex EDP tape-drive operation was moved to Southern California and combined with the acquired operations to form the Ampex Computer Products Company.

Long believed that smallness had been the key to Ampex's creativity. An important aspect of Poniatoff's style had been personal contact and rapport with the research staff. As president, Long sought to institutionalize this intimate atmosphere by decentralizing the company into five completely independent divisions, which he called the "five little Ampexes": the Professional Products Division (professional audio and video recording), Military Products, Computer Products, Orr Industries (tape manufacturing), and Ampex Audio, Inc. (consumer audio recorders). He also established Ampex International, for foreign sales and manufacturing operations.[24]

The decentralized structure, while appealing in concept, induced in practice a proliferation of administrative staff that soon proved untenable as Ampex faced its first fiscal crisis. In 1960 Ampex's VTR sales leveled off, in part because of a general recession, in part because of new competition from RCA, but also, perhaps, because of temporary market saturation. Most broadcasters already had their Quads and only new products—and the advent of color and solid-state machines—would generate further sales in the U.S. market. The company did not anticipate this drop in sales and inventories accumulated. Moreover, its plants had not always been keeping up with engineering changes. The result was $4.3 million in obsolete and excess inventory by the end of fiscal 1961. Long resigned in June 1961 after reporting a net loss of $3.9 million on $70 million in sales.

To put the company back on course, the Ampex board hired William E. Roberts, vice president of Bell and Howell, a company successful in both professional and consumer products in the field of photographic equipment. Roberts had been at Bell and Howell during most of his career, had run the company during World War II and had been its *de facto* operating head during much of the fifties. Because of competition with another vice president, he did not expect to become chief executive officer and was therefore seeking a new challenge.[25]

ROBERTS STRIVES FOR RENEWED GROWTH

Roberts eagerly took on the job of reviving the ailing Ampex. Exhibiting a degree of self-confidence that would eventually alienate some of his staff, he immediately attacked the malfunctioning organizational structure, consolidated the five divisions, established tight fiscal controls, and centralized sales operations. In the shake-out such measures required, he fired some 200 employees at all levels.

Thus, at first, Roberts trimmed the organization, focussing its efforts. But this proved to be only a tactical step. It soon became evident that his main strategic thrust would continue in the direction set by Long, broadening the scope of the company's operations. By the end of 1962, after a year on the job, Roberts began to express his own expansive intentions.

Another manager might have chosen to achieve growth through concentration on Ampex's major established businesses. Most of the revenues and substantially all of the profits were realized in two core businesses—supplying tape recorders to broadcasters and to government (mainly military) customers. In its other businesses—principally the manufacture of recording tape and of computer memory products—Ampex participated in rapidly growing markets, but enjoyed a less secure position.

The company's most profitable business at the time was the manufacture of video recorders for broadcasting. Ampex dominated the market and enjoyed gross margins as high as 80%.[26] Improved versions of the Quad and a multitude of accessories continued to be forthcoming.[27] But Government contracts were also important: The Defense Department's missile programs and the early days of the space race stimulated demand for high-performance compact recorders for video and instrumentation applications.[28] Innovation in instrumentation recording also enabled Ampex to provide a succession of improved recording media and machines to the burgeoning computer business.

Roberts chose not to rely on these few diversified businesses alone to provide a continuation of the company's growth. The decline in revenues in 1960 that had unseated his predecessor would have been a graphic demonstration of the risks inherent in such a strategy. Moreover, the profits to be made in emergent markets must have seemed hard to resist. Whatever the reasons, Roberts chose to broaden the base of the company's operations and to strive for a higher degree of diversification, while simultaneously pressing the general managers of the core businesses to deliver steady increases in revenues and profits each quarter.

In December 1962, in a talk before the New York Society of Security Analysts, Roberts made clear his own commitment to the goal of "dynamic growth." He told the analysts that his management would be ". . . concentrating all our efforts on . . . video, instrumentation, and audio recording equipment, magnetic tape and computer memory products. All," he emphasized, "offer outstanding opportunities for growth." And so they did. But the undiscriminating pursuit of so diverse a set of business opportunities implicitly contradicted the suggestion of a "concentration" of efforts. There were, of course, some common threads linking these businesses. All the products embodied some application of magnetic technology. More important, however, were the differences encountered as the company sought to compete for the allegiance of customers with quite different needs and characteristics. These diverse businesses proved to require fundamentally different skills for success.

The subsequent growth in Ampex's revenues was achieved through a combination of means. In 1964, Roberts engineered the acquisition of Mandrel Industries, a modestly-profitable producer of equipment for geophysical exploration. At the same time, he continued to encourage, wherever the opportunity arose, a characteristic Ampex strategy—invention of high performance novel equipment to meet new needs. Roberts, whose reputation at Bell and Howell suggested that he would not value research highly, explicitly declared his commitment to Ampex's traditional strategy of aggressively emphasizing research and development. In an interview a few years later, he claimed a change in attitude, attributing it to "something in the California air . . . a venturesomeness you don't find in a staid industrial spot like Chicago."[29] For example, a sophisticated and expensive information retrieval system called "Videofile" was marketed to users who needed instant access to volumes of recorded information. Championed by research director Arthur Hausman, Videofile systems were installed at NASA, the Southern Pacific Railroad Company, and the Royal Canadian Mounted Police.[30]

Beyond all this, Roberts launched a major foray into consumer markets, attempting to transfer established Ampex technologies to meet the different needs of consumers. Ampex's first consumer division, Ampex Audio, Inc., had been dissolved upon the demise of the "five little Ampexes" when Roberts first arrived. High quality audio recorders sold to affluent and discriminating consumers were subsequently manufactured under the umbrella of the professional audio products operation. In 1962, the market for a broader consumer line still seemed wide open; although the Japanese already had a large share of the lower end, the market was in its infancy. Given the rapidly improving technology, many firms, U.S. and foreign, perceived lucrative opportunities ahead. Within this context, President Roberts decided to establish Ampex "as the number one supplier of consumer audio equipment."[31]

MASS-MARKET VIDEO OPPORTUNITIES

Speculation about the potential mass-market for videorecorders had begun as soon as Ampex demonstrated the first VR-1000. One published report quoted Ampex's president, George Long, suggesting that "eventually they [VTRs] might be mass-produced for home use by persons who want to see a program over and over again or want it recorded during their absence."[32] A few days later a reporter elaborated: "A more visionary project is the thought of a home recorder and playback for taped TV pictures. Why not pick up the new full-length motion picture at the corner drugstore and then run it through one's home TV receiver?" He then cited an Ampex engineer as predicting that such a home recorder "conceivably might be ready in five years or so."[33]

Two years later a sober appraisal of the company by a respected Wall Street firm continued to reflect this enthusiasm, concluding with this statement:

> It is impossible to estimate the potential market over the long run for video tape recording equipment, so many are its applications. Video tape recorders may some day be found in offices, plants, and homes; eventually small hand cameras using magnetic tape rather than film might be used by home and professional movie makers; video tape recorders may play these films back through conventional TV sets.[34]

As these comments imply, observers quickly anticipated the several ways in which a VTR would eventually be used by consumers— convenience recording of broadcast programs; movie rentals; home movie production. Because each type of use would require a different constellation of performance features, prospective innovators had to bet on which one would prove initially to have sufficient appeal to motivate purchases of expensive equipment by large numbers of households. Also highly uncertain was the choice between two fundamentally different technical approaches to recorder design: the fixed-head technology demonstrated in the early RCA experimental machines, or a rotating-head design. Nearly two decades elapsed before these uncertainties were put to rest by the introduction of the first commercially-successful VTR for home use—the Betamax.

In retrospect we can see that the Betamax innovation required significant technical advances in several constituents of VTR design. Microelectronic circuitry and high-energy recording tape—fundamental technologies not invented until the 1960s—proved essential. At least as important were inventions in the design of scanners for rotating-head recorders, in packaging of tape in cassettes, and in the methods of manufacturing magnetic recording heads. Moreover, substantial evolutionary improvements in each of these novel technologies took place before they were synthesized to provide the cost and performance characteristics of the Betamax and succeeding consumer VCR products.

These new technologies were created by the efforts of numerous firms in several countries. Ampex was a significant contributor, but important developments came also from rivals like Sony, Philips, and Matsushita, as well as from suppliers, like the producers of recording tape and of electronic components. No single firm dominated the technological frontiers, and most of the firms striving to develop VTRs for home use were quick to assimilate technologies developed elsewhere.

The business risks and rewards inherent in the pursuit of mass markets for video recording were necessarily different from those that had faced the pioneers in broadcast recording. In the latter case, the identity of the first customers—U.S. TV networks—and their first application of the machine—for time-shift—not only were known but were also sufficient to make development of the recorder a rewarding business opportunity. In contrast, the character and magnitude of the opportunity for a consumer product were highly uncertain. Furthermore, while the broadcast innovator could gain substantial economic advantage from being first—because of the customers' need for standardization of equipment—the first firm to develop a product well-suited to home use would not necessarily have as great an advantage in garnering the rewards in the commercial marketplace.

Thus the development of a commercially-practical VTR for home use turned out to be a very different sort of task from that of creating a machine for broadcasters. The Ampex team led by Charles Ginsburg had perceived itself to be in a race to invent a VTR for broadcasters. That race turned out to be a sprint, and the winner got to keep the gold. Rivalry to open the market for mass-produced machines turned out to be more like a marathon. Moreover, while the winner might earn acclaim, the end of the race to innovate merely signalled the beginning of rivalry for market share and profits. To profit from this innovation, firms needed to have significant marketing resources and the payoff has come as much from how they managed commercialization and improvement of the innovation as from how they created it.

Ampex Corporation in the 1960s was ill-suited to that sort of contest. The company had prospered largely from engineering achievements, not from advantage won in the marketplace. The rewards of a dramatic achievement like the VR-1000 were vivid in the consciousness of key managers and engineers. In its approach to other opportunities, like the home VTR, the company sought similarly to gain a decisive technical advantage. While the considerable skills of its engineering staff made it a formidable competitor in every "sprint," Ampex was ill-equipped to sustain a marathon, let alone to compete for the real rewards after the technical race had been settled.

Technical Foundations of Mass-Market Video

While journalists and investors felt free to speculate about the manifold potential markets for video recorders, engineers were very aware of the technical obstacles that had to be surmounted before any new market could be tapped. The VR-1000 was a great technical achievement, creating a revolution in broadcasting. But the Quad format, with four heads creating 16 discrete magnetic records for each television field, made it inherently expensive to build and demanding to operate and maintain. In broadcasting the economic rewards were tangible and the studios staffed to support such equipment; elsewhere there were few ready buyers for that sort of equipment.

A different kind of scanner would be required to create a VTR that could succeed commercially in other potential markets. It was natural to experiment further with fixed-head and helical scanner designs— concepts which had been explored in several firms during the 1950s.

The technical challenge of developing a practical recorder based on one of these alternative scanners was still great. But the task was not as daunting after 1956 as it had been before Ampex's landmark innovation. Some of the design solutions incorporated in the VR-1000, such as the FM circuitry to encode the signal, were transferable to other types of VTRs. Good quality recording tape, a scarce commodity before 1956, became available in commercial markets. Growth in demand for recording heads suitable for video stimulated vendors to develop better materials. Ampex and others gained valuable experience in the fabrication of heads in commercial quantities.

Experimentation with helical scan concepts had been initiated independently in several widely-dispersed laboratories in the 1950s. (Figure 3 illustrates the early designs that were patented.)[35] The first working model of a video recorder using a helical scanner was built at Ampex Corporation in 1956 by Alex Maxey. As noted earlier, Maxey contributed significant elements of the scanner design for the VR--1000. With the VTR design firming up, the restless and inquisitive Maxey began to seek new challenges. As he later said, his "long suit" was not refining designs for production but generating new ones. By

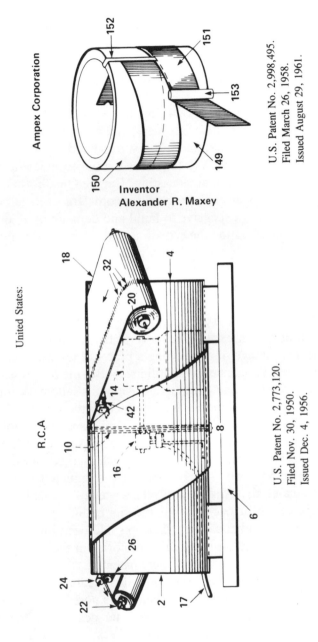

Ampex Corporation

152

151

153

150

149

Inventor
Alexander R. Maxey

U.S. Patent No. 2,998,495.
Filed March 26, 1958.
Issued August 29, 1961.

United States:

R.C.A

18

32

20

4

14

42

10

16

8

26

24

22

2

17

6

U.S. Patent No. 2,773,120.
Filed Nov. 30, 1950.
Issued Dec. 4, 1956.

Figure 3. Early Helical Scanner Designs

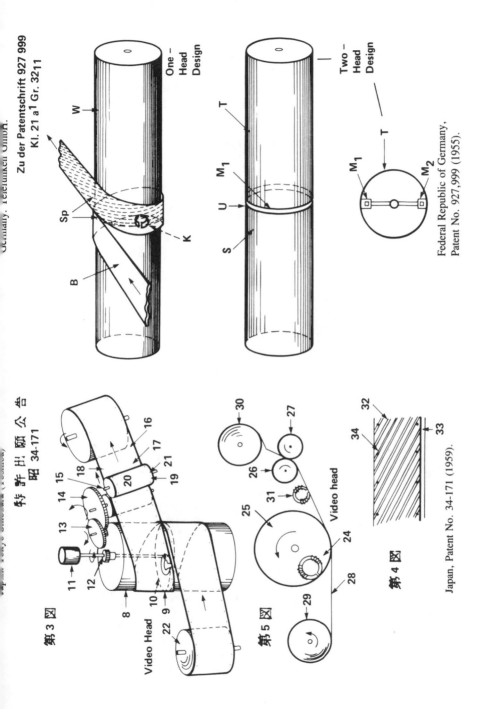

Germany. Telefunken GmbH.

Zu der Patentschrift 927 999
Kl. 21 a¹ Gr. 32¹¹

W

Sp

B

K

One –
Head
Design

T

M₁

U

S

T

T

Two –
Head
Design

M₁

M₂

Federal Republic of Germany,
Patent No. 927,999 (1955).

特許出願公告
昭 34-171

第 3 図

Video Head

11
12
13
14
15
16
17
18
19
20
21
22
8
9
10

第 5 図

25
26
27
30
31
24
28
29
Video head

第 4 図

32
33
34

Japan, Patent No. 34-171 (1959).

the end of 1955 he had built a crude prototype of a recorder using a single rotating head. Encouraged by the results, he went on to design a machine using a helical wrap. Then he followed with a series of prototypes embodying successive solutions to problems of tape tension, tape movement, and so forth. (see Figure 4 and Appendix A.)

Figure 4.   A. Alex Maxey with an early helical prototype,
1956 (Ampex archives).

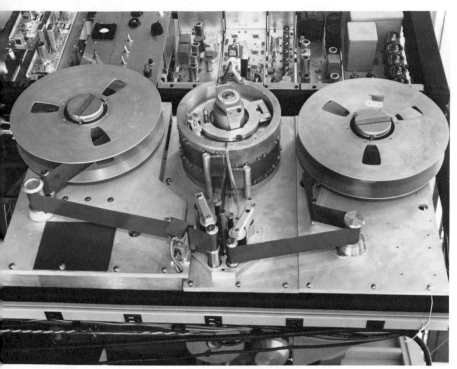

Figure 4.   B. A close-up of the tape transport of another helical
prototype, 1959 (Ampex archives).

In Japan, working from an Ampex VR-1000 purchased by NHK (the
state broadcasting company), several electronics manufacturers ''reverse engineered'' replicas that were suitable for broadcast use. Some
of these companies, including Toshiba, Matsushita, and Sony, also
adapted aspects of the design to create recorders based on helical
scanning formats.

The development effort at Ampex was not limited to Maxey's experimentation and invention. By early 1960 its engineers had built two
working prototypes of a prospective product for broadcasters. Not
wanting to reduce buyers' interest in the existing product line when the
prospective new technology was still in a primitive state, Ampex kept
its helical efforts under wraps.[36]

Nevertheless, rumors that RCA and perhaps some Japanese companies were about to demonstrate models using helical scanners
prompted Ampex to send two prototypes to the convention of the
National Association of Broadcasters in March 1960—but in secret, to

Figure 4. C. One of the ''NAB'' machines, 1960
(Ampex archives).

Figure 4.   D. A close-up of the tape transport of the
NAB machine, 1960 (Ampex archives).

be shown only if another company did likewise. The "NAB machines"—one color, one monochrome—were not really ready to be
shown; nor was Ampex ready to meet the expectations such a demonstration would have generated among its customers.[37] Meanwhile,
there was controversy within the company over whether the helical
scanner could perform at the level required for broadcasting.

In early 1961, Ampex unveiled its first helical product, the
VR-8000. Featured in the company's Annual Report as "the newest
Ampex visual memory device" (see Figure 5), the machine incorporated several features of the VR-1000. With most of the electronics
using new transistor circuitry, it was more reliable, smaller, and only
half the price of the Quad, but it did not perform nearly as well. Only a

Visual Memory:

The newest Ampex visual memory device, left, is the VR-8000 closed-circuit VIDEOTAPE recorder, utilizing one greatly simplified video recording head instead of the 4-head assembly necessary in the broadcast VIDEOTAPE recorder. Contained in a single console-type cabinet, the VR-8000 is simple, both to operate and maintain, and costs about 50% less than the broadcast recorder. It operates at one-half the speed of a broadcast recorder, thus doubling the amount of recorded material which may be stored on a reel of tape.

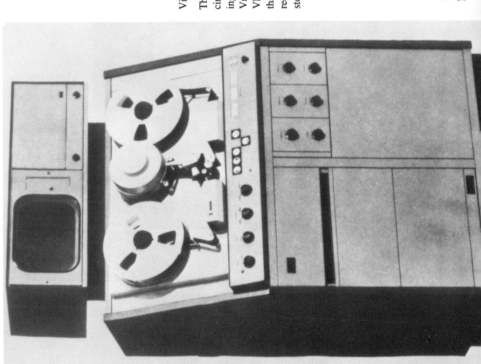

Figure 5. "Visual Memory"—The VR-8000 as featured in Ampex's Annual Report of 1961.

few reached the marketplace. Ampex engineers argued among themselves about the future of the VR-8000 design; Maxey, for example, later said he had always believed that the objective of designs using the new helical technology should not be to "out-Quad the Quad," but to do something entirely different.

The first commercially successful video recorder with helical scan was the Ampex VR-1500, brought to market in 1962. Its most innovative technical constituents were adopted directly from Maxey's experiments of the late fifties. As early as 1957 Maxey had begun working on elements of a "simple-minded machine for a simple-minded market," a model he called "Junior." Its writing speed of 641 inches per second, tape speed of five inches per second, and narrower track meant a significant saving over the Quad in the quantity of recording medium used. These characteristics, embodied in a commercial product developed under the project management of John Streets, yielded a machine suitable for "closed-circuit" (i.e., non-broadcast) television at one-quarter the price of broadcast-quality equipment.[38]

Prior to the successful introduction of the VR-1500 in 1962, Ampex had filed for more than a dozen patents on Maxey's scanner inventions. But there were still problems to be solved. Wear on the tape was one—some better way of reducing friction between the tape and the drum had to be found. Too many moving parts that had to stay properly aligned was another—the challenge here was to simplify the scanning mechanism still further. Maxey and Streets would address all these problems with a new design reduced to practice in 1964.

This new design, characterized by a self-lubricating, air-bearing scanning assembly (or "two-section scanner," as it has been dubbed), became a key part of Ampex's one-inch closed-circuit machines of the 1960s and of many subsequent products by Ampex and its competitors. Its two most important features greatly simplified helical technology. First, its scanning mechanism incorporated heads (or one head, as in this instance) attached directly to the upper section of the mandrel, now in effect a scanning drum, rather than to a separate device rotating between two stationary mandrel sections. Second, as this upper mandrel section rotated, it generated an air film that supported the tape.[39]

Maxey recalled that his aim had been to develop a machine, "elegant in its simplicity," that could perform well enough to serve customers who needed good quality video but could not afford machines built on the Quad format. But the advances embodied in this new design were a significant step toward making low-cost helical VTRs practical for home and industrial use, and would later be incorporated (under license from Ampex) not only in the broadcast "C" format made by Ampex, Sony, and other companies, but in most modern helical machines—including the U-Matic, VHS, and machines made by Philips, Matsushita, and Grundig. In 1965, just after Maxey left Ampex to form his own company, the development of products based on this design was taken up in another part of Ampex, under different management, by the recently established Consumer and Educational Products Division in Chicago.[40]

THE BIRTH OF A NEW INDUSTRY

When a company undertakes to create and market a radically new product, it faces two broad classes of choices. One set determines the definition of the product, broadly and in detail. The second set of choices determines the character of the organization that will create and exploit the innovation. A small firm can focus its energies on the definition of the product; its organizational options are limited. The large organization can afford to hedge its bets on the choice of product technology and features, but it must contend also with a wider range of organizational possibilities.

When the top management of Ampex addressed the opportunity of marketing a video recorder for consumers in the mid-1960s, several organizational solutions were possible. The Professional Products Division, based in Redwood City, had developed and was marketing Ampex's first low-cost recorder, the VR-1500. It also provided the framework for the development of advanced scanner designs, being led by Maxey and Streets. But the part of the company already engaged in consumer markets was in Chicago, making and selling inexpensive audio recorders.

In 1952, when Ampex was small, such questions did not arise. Alex Poniatoff had set the charter for Ginsburg and the small group that created the first VTR, sheltering them from many of the pressures of the rest of the organization. By 1964, the company was ten times larger and Poniatoff was no longer directly involved in operations. The Chief Executive who shaped Ampex's home video ventures, William Roberts, was of a different temper, and proved to be poorly suited to the entrepreneurial role that circumstances demanded.

The result was rivalry for the "franchise" to develop a home VTR, and controversy surrounding key choices in the definition of the product. The visible manifestation of this was the introduction of two different home recorder products by two different organizational units, within the first six months of 1965. In February, under the auspices of Vice-President Gus Grant's Professional Products Division, the company introduced the VR-303, a fixed-head (longitudinal scan) recorder. In June, a wholly different design, developed by the engineers of the consumer products business in Elk Grove Village, was demonstrated at the Consumer Electronics Show (then called the Music Show) in Chicago. Identified as the VR-7000, this machine used a rotating head in a helical scan format. Neither product found a niche in the consumer market. The fixed-head technology of the VR-303 was soon perceived as a "dead-end" and the venture terminated (with a $1.2 million accounting write-off). The VR-7000 survived, repositioned as an educational and industrial product, and gave birth to a new segment of Ampex's video business. It led to the creation of a business unit focussed on educational and industrial video, thus establishing the foundations on which Ampex would try, later in the decade, to claim a share of the mass market for VTRs expected to emerge in the 1970s.

The Division headquartered in Elk Grove Village had been established as a direct outgrowth of Roberts's decision in 1962 to commit the company to achieve leadership in consumer audio markets. Given the influx of high-technology industries, the cost of labor and materials in California was a handicap for a business in which manufacturing costs were a primary competitive factor. Hence the decision to move Ampex's consumer business away from California to an area where labor costs were lower, suppliers available, and parts not so expensive: suburban Chicago. Roberts and several colleagues had come to Ampex from Chicago and knew the area. Television set manufacturers were concentrated there, as were the low-cost suppliers needed for the manufacture of consumer audio recorders. In addition, a location in the center of the continent would facilitate national distribution of products for a mass market. Thus did Ampex create its first organization outside the Bay Area with a fully-integrated engineering, manufacturing, and marketing capability.

In September 1963, Roberts sent a corporate vice-president, John Latter, to Chicago to start up the new division and to oversee the construction of a plant in the suburb of Elk Grove Village. Latter had been with Roberts at Bell and Howell and had come to Ampex in 1961 as corporate controller. As chief engineer Roberts chose Rein Narma, who had joined the company in 1959 as chief engineer for the unit making audio recorders for professional markets.[41] John Trux, who had several years' experience in advertising and marketing for consumer products firms in the Chicago area, including Bell and Howell, was hired as marketing manager. A skeletal group worked out of a rented store until their new audio plant was ready in December.

Anticipating this move, Narma earlier had asked two engineers in the audio division, Austin Ellmore and Ray Siebert, to design a product suitable for the consumer market.[42] The two were an effective team that became particularly expert in designing tape transports. But when the time came to move the company's consumer operations to Chicago, Ellmore and Siebert did not wish to leave sunny California. The company permitted them to set up shop in nearby Los Gatos, where they continued to work for Narma long-distance.

Someone at the plant in Chicago had to turn Siebert's and Ellmore's design into a manufacturable product. That responsibility went to Delmar Johnson, a jack-of-all engineering trades who had never seen a tape recorder before coming to Ampex. His previous job with a government contracting company, however, had given him broad experience in a wide variety of engineering tasks. Hence his assignment when he arrived in Elk Grove in March 1964 was to debug the Division's first product.

Within a year, Elk Grove audio products, rapidly expanding into a full line, appeared to be launched successfully in the marketplace. Ampex claimed to have captured 50% of the high-end market and made significant inroads into the middle range. Trux quickly developed a retail sales network of some 800 outlets throughout the country.[43] Competition was still moderate. Foreign imports, mostly from Japan, accounted for almost all sales of audio recorders under $50 and about half the market for those costing $50 to $100, but they had taken only 30% of the market for recorders over $100, where Ampex was strongest.[44]

While Elk Grove was forging ahead with its audio products, the magnetic recording industry was shaken by the first public demonstration of a low-cost video recorder aimed at the consumer market. It took place in December 1963, under the auspices of Cinerama, Inc. The product was a fixed head (longitudinal format) design developed in England by a company called Telcan.[45] Although the picture quality was generally regarded as inadequate and highly erratic, the indicated retail price of $175 attracted considerable media attention.

Renewed attention to low-cost video evoked new initiatives at Ampex. Although the core of the engineering group at Ampex who had worked on the quad and the new helical formats remained skeptical of any attempt at developing a longitudinal video recorder, other people were willing to give it a try. Vice President Gus Grant, head of marketing and later General Manager for Video and Instrumentation Products, became the champion for development of the fixed-head scanner that provided the basis of the abortive VR-303. Roberts, having already committed the company to consumer audio and clearly eager to expand Ampex's efforts in this direction, was responsive, in turn, to each of the engineering enthusiasts seeking to commercialize technology that would lead to truly low-cost video recording. He had been at Bell and Howell when that firm had prospered from the introduction of 8 mm home movie equipment and seemed to some to be beguiled by the prospect of an electronic substitute that would challenge his former employer's business.

Narma and his colleagues in Elk Grove Village were quick to claim their own right to the franchise for the consumer video business. Narma had left the California operations reluctantly, but had been persuaded by Arthur Hausman, the chief technical officer, that there was little future for audio operations on the West Coast. He recalls that Hausman was optimistic then about the prospects for the evolving technology of helical scanners in "semi-professional" applications. Hausman assured him that low-cost video products would be assigned to Elk Grove Village when they were ready.

Narma moved quickly to keep his team in the contest. After Ellmore and Siebert at Los Gatos had forwarded their audio designs to manufacturing engineers at Elk Grove Village, Narma sought to exploit their expertise in design of tape transports, assigning them to design a machine that could handle the high speeds necessary to record video signals in a fixed head format. Their prototype, Johnson later observed, worked as well as the more publicized VR-303. Narma also asked Johnson, once the audio recorder went into production, to develop a longitudinal model. Doing so convinced Johnson that the problems with longitudinal recording were immense. He was saved from having to solve them by news from Redwood City in November about an air-bearing scanner that would make single-head helical recording practical. This was the two-section scanner developed by Alex Maxey and John Streets.

Late in November, soon after hearing about Maxey's innovation, Narma and three of his engineers travelled to California to see it.[46] After seeing a breadboard version of the scanner, the visitors decided they could make one too. Narma gathered some parts—drive motors, VR-1500 printed circuits, and so forth—into his suitcase and the foursome flew back to Chicago.

Johnson and Boylan, with Narma's eager support, set to work. A mere eleven days later they had a picture that was "far better than that of the longitudinal models." After refining their prototype, they flew back to California to show it off. Those who saw it—including Poniatoff—seemed astounded by the performance of the small, and apparently inexpensive, machine. They readily gained backing for their proposal to enter the home video business.[47]

Once again, Ampex found itself in a race to innovate in video recording. This time the most serious competitor was from Japan, the Sony Corporation. Sony had demonstrated its first helical design in January 1961 but never carried it to commercialization. In 1962, soon after Ampex introduced the two-inch VR-1500, Sony demonstrated the PV-100, a two-inch machine that it sold in small quantities.[48] As early as 1962 Sony's advanced development group had begun to work on a one-inch format, but switched in 1964 to a two-head, half-inch design that became the CV-2000. This monochrome recorder, which Ampex viewed as simply a scaled-down version of its own VR-1500, was first shown in Japan on November 17, 1964, and in New York City on June 2, 1965.

Holding the baton for Ampex in this race were Narma and his associates. Emboldened by their rapid progress in embodying the Maxey-Streets scanner innovation in a working model, they committed themselves to build a prototype for public demonstration at the next Consumer Electronics Show, a few months away.[49] That period is remembered as a remarkable time, when it wasn't unusual to find the entire engineering team still at work after midnight. To some it may have seemed reminiscent of the frantic period just before the VR-1000 was first unveiled.

Narma's team had to make a series of difficult and fateful technical choices while contending with vexing organizational dilemmas.

To begin with, they were trying to create a reliable, low-cost consumer product from a novel technology about which no organization had any great depth of knowledge. Ampex's existing product with a helical scanner, the VR-1500, required frequent service to keep it operating. The fabrication of recording heads in large quantities to meet demanding specifications was a new challenge.

The team in Chicago was determined to show that they could meet those technical challenges, and were proud when they designed and built the first prototype without transferring a single engineer from Redwood City. At the same time, Narma recognized that they were unavoidably dependent on the expertise of Ampex's experienced video engineers. As he later observed, real know-how in video recording existed only in Redwood City, and perhaps in Tokyo, at that time.

Interaction with Redwood City was a source of conflicting pressures. Within the video operation, some had believed all along in the potential of the helical scanner and thought the responsibility for commercializing it should have remained with them. Others seemed to want to keep development on a path that would not lead to machines competitive with the Quad. At headquarters, meanwhile, Roberts pressed for rapid progress to stake out the consumer market.

The most critical technical issues pertained to the configuration of the scanner, which established the "format" of recording. Ampex, unlike most of its rivals, sought to establish a format for helical scanners that would endure, as the Quad format had. This led to a set of technical choices (writing speed, for example) creating a format robust enough to be incorporated in high-performance machines, yet capable of manipulation (e.g., a half-speed design) to suite low-price market segments. Also critical was the decision to use a single recording head, rather than two, as were used in the VR-1500 and in Sony's new design.[50]

Ampex engineers were convinced at that time that a low-cost recorder must use only a single recording head. First, the heads themselves were inherently expensive. Second, the need to locate dual heads precisely in the same plane created difficult mechanical problems in volume manufacture.

The art of manufacturing recording heads for video was itself poorly understood in the mid-sixties. Narma was startled when the first quotes (from Ampex's video head department) for heads for his new machine indicated a unit cost above $100. He turned instead to the Sunnyvale operation that had been producing heads for his stereo recorders and they soon met his needs. Designing heads with the prospect of volume production in mind, they soon achieved the ability to produce at a rate of several hundred units per month at costs below $10 per unit.

The net result of these choices was that Ampex developed a machine that produced high-quality recordings yet could plausibly be offered at a price competitive with Sony's prospective offering.[51]

The team in Elk Grove Village succeeded in producing a prototype recorder which they could demonstrate at the Chicago Music Show (now called the Consumer Electronics Show) in late June. Here the Ampex group would meet its competition for the first time and have a chance to gauge the reaction of both the industry and the public to its product.

The appearance at the show of Ampex's new product, dubbed the VR-7000, along with Sony's low-priced machine, attracted attention.[52] The enthusiastic response in the trade and popular press seemed to confirm the faith of those who believed in a latent consumer market for video recorders. Products of both companies were featured in a *Life Magazine* story in September 1965, headed: "A new pastime with a big future: Tape-It-Yourself TV." Photos showed Ampex and Sony VTRs alongside backyard swimming pools, recording pictures from small video camers. The text was exhuberant, suggesting that "it's going to be the Polaroid of TV and home movies" (see Figure 6.)[53] It went on to say that "the basic use of these units will be to record TV programs. . . . This would not be cheap since an hour's tape now costs up to $60. But there's already talk about a time when taped recordings of Broadway musicals and plays will be made and either sold or rented." *Merchandising Week* quoted an enthusiastic sales pitch: "Take a piece of paper and start writing; by tonight you will have thought of 2000 uses for a video tape recorder."[54]

At Elk Grove Village, work progressed on the VR-7000. By November 1965, when the operation moved to its new plant, the engineers had completed several variations in six prototypes for sales demonstrations and further development. At Redwood City's behest, one of these models was packaged in a walnut case, equipped with a control center and video tuner, and dubbed the VR-6275. This machine was to be the basis of a full line of home video recorders, from a table-top recorder/player for $1,095 to a complete system with video camera and receiver for $2,495.[55] In December 1965, *Financial World* reported that Roberts expected these recorders would be ready for shipment "next month"; the same article cited Roberts' admission that

A new pastime with a big future
Tape-It-Yourself TV

Figure 6. Life Magazine Feature Story, 1965

"initially" Ampex's product would not be "in the most popular price category" but would be the "Cadillac" of the market.[56]

While preparations for production of the new machines went ahead, Latter sent his marketing manager, John Trux, to Hawaii with a prototype for a market test. Ampex's consumer audio business had been successful there, and Hawaii—"off the beaten track"—was a safe place for a risky test. With the help of a public relations firm, Trux gave four or five presentations daily to 40 or 50 people at a time, then distributed questionnaires that attempted to find out whether the audience would buy such a product and at what price. The results were sobering. Trux found no consumer interest in the machine at a price at which it could be profitably sold.[57] Not only was it too expensive at $1495, but it recorded only black and white pictures when color television was just taking off; moreover, it was complicated to use. Roberts, who had been convinced that video recording would replace 8mm home movies, was upset by the results of the market test. The Elk Grove group simply concluded that the home market required color, and that development of a color camera that would be inexpensive and easy to use was beyond the reach of current technology. But the news from Hawaii was not all bad: while the questionnaires revealed a lack of interest in Ampex's VTR for the home, they did exhibit an enthusiasm for the machines in other applications.

Production of the Ampex VR-7000 began early in 1966 and 44 units had been shipped by the end of March. Despite the results of the market test, Roberts was still hoping to prove that a consumer market was indeed waiting for Ampex's product. In March 1966 *Business Week* quoted Roberts' anticipation of $10 million in sales of home video recorders the first year and $100 million by 1970.[58] Two months later, Ampex's Annual Report featured the home video units, proclaiming their ability "to store and repeat outstanding television programs, to make instant home movies of family events or dad's golf swing or to play prerecorded tapes of outstanding dramatic works"— in short, to offer something for everyone. The report stressed the lower cost and reusability of videotape as important advantages over 8mm home movie film and touted the pleasures of being able to record a World Series game for viewing at one's leisure. The price with walnut case was $1,495 without a TV receiver; the camera was $529 extra (see Figure 7).

Six manufacturers of videorecorders displayed their wares at the Chicago Music Show in July 1966. Among them was Matsushita,

Figure 7. Ampex VR-6275

which introduced a low-priced consumer model, like Sony's a half-inch format with two recording heads, and showed it under the Panasonic and Concord labels. At the same time, despite the failure of several fixed-head machines—including Telcan's—several European and North American firms announced plans to introduce such models at startlingly low prices.

Yet none of the so-called home video recorders of the mid-1960s succeeded in the marketplace. Not Ampex, not Sony, nor any of the others. During 1966, both pioneering companies, Ampex and Sony, drew the same conclusions. First, the products they were offering to consumers inspired very little consumer demand. Second, the clues about what consumers would desire in a VTR indicated the need for significant advance in the technology. Third, fortunately, other customers were interested in buying the available VTR products: schools, hospitals, offices, factories, and the like. And thus was born the "audio-visual" (A-V) segment of the video recording market.

Opening Up the Audio-Visual Market

To sell to users in schools and colleges, hospitals, factories, and the like, Ampex had to develop a broad line of products and accessories to suit varying needs, and also to create a strong marketing support program.[59] The engineering group at Elk Grove Village plunged into video, sharply cutting back their innovative efforts in audio. A new tape duplicating center provided high speed quantity duplication of tapes and transfer of recordings from one format to another. A training center at Elk Grove Village, called the Ampex Video Institute, offered instruction in closed-circuit television techniques to Ampex staff, dealers, and customers.[60]

Sales were more than encouraging; about 2500 units of the first products—models VR-6000 and VR-7000—were shipped to dealers in 1966. By the end of the year shipments were running at nearly 400 per month. As Jack Trux later recalled, the line was so successful that Ampex had to "invent accessories overnight."[61]

But the euphoria induced by the rush of orders in 1966 proved to be short-lived. The rate of shipments dropped sharply in 1967 and by mid-year was only half what it had been at the end of 1966. The management of the Consumer and Educational Products Division [CEPD—renamed to reflect its broader charter] found itself, once again, battling on three fronts. Its engineers labored to master an emerging technology sufficiently well to be able to offer an appropriate line of products at competitive prices with adequate reliability of performance. Second, the Division had to mobilize its marketing resources in order to win sales against a growing international group of competitors.

While the technological and competitive challenges must have looked similar to Ampex's rivals, the management at Elk Grove Village had to contend with a "third front," as they encountered conflicting pressures from Redwood City. While some of the video professionals at Redwood City pressed them to enhance the performance of the technology, others resisted the development of products that encroached on their market for high-performance Quad recorders. Neither the technology nor the market were ready for the creation of a true mass-market business, but Roberts kept pressing them to go after the consumer.

The effectiveness of the Elk Grove Village engineering effort remains a matter of controversy. When they were in good operating condition, the CEPD video recorders worked well. Picture quality was excellent, and the products had other high-performance features, including interchangability of tapes, not only among units of the same model, but also between different models in the line. Thus a school could make a tape on one of the top-of-the-line 7000 series, duplicate it, and have copies played on the less expensive VR-6000.

The poor reliability of the equipment in the hands of users, however, was a continuing source of difficulty. The division worked hard to train its dealers' service personnel. Some of the managers then responsible argue that the rate of failure was no greater than was normal at that time in consumer electronics products. Others recall that the products experienced severe reliability problems, both in mechanical design and in manufacturing quality control, to the extent that Ampex "opened the door" to Japanese competitors.[62]

In mid-1967, CEPD hired a marketing consultant to gather information about the competitive scene. According to their estimates, sales of videorecorders for audio-visual applications in the U.S. had reached a cumulative total of $15 million (at factory prices) by mid-1967—still a small market.[63] Machines using one-inch tape accounted for 70% of the dollar value of sales, while the less-expensive half-inch models represented a majority of units sold. Ampex was the dominant producer in the one-inch segment, with more than 50% of the total market (in dollars). In education it was particularly strong, with 56% of the units, and 68% of the revenues. Its share of the medical market was also high. Competitive products using one-inch tape were offered by International Video Corporation (IVC),[64] and several Japanese producers, with Philips of Holland expected to enter the market soon.

Sony and several other Japanese producers offered machines using half-inch tape with formats originally developed for the envisioned consumer market. Sony was quick to broaden its line to suit the requirements of the A-V market. In July 1966, they demonstrated a prototype of a battery-operated portable recorder using the CV-2000 format.[65] In 1967 Sony began to produce a one-inch helical recorder offering higher picture quality, the EV 200 series. By the end of 1967 they were producing at least 15 different models using the 1/2" and 1" formats.[66]

The Japanese competitors were alone in offering the option of half-inch formats. In general, these machines provided somewhat lower picture quality, but offered simplicity of design, ease of operation, and much lower cost. Users found that they were reliable in the hands of unskilled operators and needed little maintenance.[67]

Ampex's consultants, in commenting on these Japanese designs, noted that the industrial users seemed particularly oriented toward price, reliability, and simplicity of design, and did not need the higher performance capabilities of one-inch formats. In the industrial segment, they estimated that Ampex was getting only about 40% of the market in revenues, and about half that share in units.

At first, CEPD chose to compete in a manner characteristic of Ampex strategy in other markets: they tried to fence off the high end of the market and emphasized their use of a single format in a market in which incompatible formats were proliferating.

The first step up in the product line was a recorder for color video, the VR-7500, shown in the spring of 1967 at the NAB Convention and first shipped in mid-year.[68]

That the VR-7500 was introduced at the broadcasters' convention was indicative of what was happening to Elk Grove Village's video products. As CEPD responded to competitors, moving to higher performance products as a way to differentiate their offerings, they inevitably narrowed the distinction between their line and the products offered to broadcasters from Redwood City.[69]

Of the conflicting forces influencing management at Elk Grove Village, then, the discipline of a competitive market seems to have been most influential at first. CEPD came to offer a broad line of products designed around a common format. Through higher performance (in terms of picture quality), interchangeability of tapes, and effective marketing support, they sought to make the Ampex one-inch standard competitive in a variety of markets, from low-end closed-circuit to broadcasting.

Pressure on the Bottom Line

Top management, back in Redwood City, defined the important financial constraints governing CEPD's ability to compete in the emerging video marketplace. Roberts pressed continually (as he did with every profit center manager) to obtain increases in reported earnings. At the same time, he restricted CEPD's investments to the level they could fund out of their own operating cash flows.

During the mid-60s the Ampex Corporation found itself chronically short of cash and increasingly turned to lenders to meet its substantial investment requirements. Spending for research and development, the *sine qua non* for a high technology company, was constrained, company-wide, despite Roberts's public assertions to the contrary. R & D expense as a percentage of sales was only about 6% after fiscal 1964—down from more than 8% during the late 1950s, a level consistently maintained by Hewlett-Packard and other high technology firms. (see Figure 8.) The Elk Grove effort could not compete with other projects for the limited funds available. Videofile alone, for example, had gobbled up $6.5 million in research and development funds by 1968.

The financial pressures on CEPD grew more intense in late 1967 as corporate profits declined for the first time since Roberts took over. For the fiscal year ended April 30, 1968, Ampex Corporation reported a decline of one-third in net profits compared to the previous year. As top management began to reassess its least profitable operations, Elk Grove came under severe scrutiny.

CEPD's video business, facing growing competition from Sony and other Japanese producers, seemed unable to regain the higher volume of 1966 and unable to move into the black at its current rate of sales. At the same time, the division's consumer audio business was beginning to suffer from the onslaught of Japanese competition and the lack of further innovation in the Ampex line. Low-cost, high-quality audio

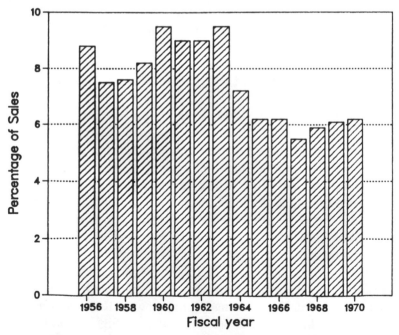

Figure 8.   Research and Development Expenses

recorders from Japan were attractive to all sorts of buyers, and Ampex equipment became less competitive even among the sophisticated ones. Ampex anticipated the popularity of the new audio cassette introduced in 1967 by Philips of the Netherlands; later that year it was the first American company to import and sell audio cassette recorders. But the Philips-made recorders quickly lost their ability to compete on the market, as the Japanese began to produce the equivalent product much more cheaply. Ampex soon bought these inexpensive cassette machines from Japanese companies to fill out its line; nevertheless, it—like many U.S. companies—was never able to compete with the Japanese labels in the low and medium price products that took over the market.

At year end, Latter and Trux left Ampex, returning to Bell and Howell. In an internal memorandum announcing that Rein Narma would succeed Jack Latter as General Manager of the Consumer and Educational Products Division, as of January 15, 1968, Roberts noted that Latter had resigned because of a "conflict in operating philosophy."

At first, no general manager was assigned to the Elk Grove video business. Narma, who had helped nurture it into being, sought capital to develop what he viewed as a business with a bright future. But top management was still more concerned with current results and unwilling to fund Narma's requests.

Taking Aim at the Home Market Again

Introduction of an inexpensive model, the VR-5000, gave a boost to CEPD's sales in mid-1968. Narma assigned Austin Ellmore's advanced development group in Los Gatos to develop an even smaller, simpler, and cheaper machine, still using a one-inch format. This was to be called the VR-7700. But concern about CEPD's competitive strategy persisted in Redwood City. In 1967 the Division's consultants had observed that "CEPD needed to develop a long range marketing plan for its VTR products to ensure that its daily activities continued to be consistent with and helpful to the attainment of its longer term growth."[70] Narma, like his predecessor, found it difficult to meet his bosses' demands for better short-term performance while also pursuing a plan geared plausibly to longer-term growth. Roberts continued to press for both. In mid-1968, seeking better intelligence on the market prospects for video, Roberts sent a personal emissary to look over the business. The person he chose was Richard J. Elkus, Jr., son of one of the more senior and influential members of the Ampex Board of Directors.

The younger Elkus had joined Ampex in 1959, held a series of middle-management jobs, and left after 5 years to form his own company.[71] In April 1968 he rejoined the company as assistant to the president. Young and ambitious, he threw himself into his first assignments. After a quick study of the market for disc packs, he was asked to look at the home video market and to review Ampex's helical video recorder operations at Elk Grove Village.

Working from a base in Redwood City, by year-end Elkus had produced a persuasive report calling for a major shift in video product development strategy. As he later recalled, "I wanted to develop a strategy which allowed Ampex to provide a product which served a market that did not require a massive investment in distribution or a massively large, efficient manufacturing system competitive with the Japanese." Arguing that Ampex should do something that it could do better than anyone else, Elkus thoroughly investigated all aspects of its business—technological resources, problems of manufacturing, and marketing strategies.[72] He concluded that Ampex had unmatched technological depth in magnetic recording, but was weak in manufacturing and distribution. To understand and anticipate the needs of users, he interviewed people in a major university, a large industrial firm, and a large insurance company. He also drew further information from the earlier work of a consulting firm.

In a detailed and insightful report issued November 12, 1968, Elkus analyzed market needs, technology, manufacturing costs, and competition; he concluded that Ampex should compete in the growing industrial market for video recorders, and proposed in detail a design and strategy to do so successfully. Moreover, he explicitly stated that Ampex belonged in home video recording and made the home market an integral part of his proposal.[73] Elkus's review gave careful attention to the design of the proposed VR-7700, under development at Los Gatos, but his analysis of market needs and competitive offerings led him to conclude that it should be scrapped in favor of an entirely new concept.

Analysis of the A-V Market

Elkus segmented the audio-visual market according to type of use rather than user, identifying three categories: (1) repetitive viewing, (2) semi-repetitive viewing, and (3) instantaneous response. "Repetitive viewing'' involved material of sufficient importance that professional production techniques—often in a studio setting—would be employed to make a master tape, from which copies would be prepared for distribution and viewing in the field. For example, a large airline might produce in its own studio a high-quality, one-inch tape for maintenance training; copies for distribution throughout the company could then be made on one-half-inch tape for playback on less expensive, smaller machines.

''Instantaneous response'' referred to the recording of events for immediate playback to participants or others in the same location. This market wanted reliability, portability, ease of operation, and economy in machines that would be used by unskilled operators concerned with the immediacy and convenience of the presentation. Applications were numerous—tutorial situations, role-playing in education, observation of group processes, interviews. Semi-repetitive viewing involved applications reflecting a mixture of uses, with some programs produced in studios with high-performance equipment and others generated in the field by persons relatively unskilled in video.[74]

Elkus emphasized that the different formats chosen by Ampex and its Japanese competitors expressed ''basic statements [of] marketing philosophy . . . in terms of performance, reliability, [and] interchangeability. . . .''[75] The Ampex line appeared to have been designed to meet the needs of the semi-repetitive viewing segment of the market as well as the production portion of the repetitive viewing market. The instantaneous response segment (and the playback portion of the repetitive viewing segment), Elkus claimed, would prefer simple, low-cost designs, reliable and easy to use, even though such machines offered slightly inferior picture quality. Moreover, it was the latter segments that seemed to be growing most rapidly. For them, the Japanese half-inch formats were well-suited.[76]

Implicit in Elkus's analysis was the suggestion that video was not only invading the ''16 mm market'' for A-V applications, but also beginning to spawn an entirely new market, ''instantaneous response,'' for which conventional photographic techniques were inherently unsuitable. The central conclusion of his report was the explicit argument that, for this use, customers required a recorder different in concept from conventional videotape recorders. He urged that Ampex develop ''a device which the customer may use as 'an audio visual scratch pad'.''[77] This would be a market segment unmet by competitors' products. Having seen the Japanese companies take over the audio recorder market, Elkus did not want to confront them head-on.

Elkus offered a detailed analysis of competitive machines, with an appendix listing specifications and features of 10 models produced by six Japanese firms, Philips, and IVC. He showed how various manufacturers of half-inch machines had given differing priorities to picture quality, reliability, and other characteristics when making basic design tradeoffs. Sony, for example, had employed a skip-field technique in the CV-2000 to permit broad recording tracks offering greater tracking reliability, though lower picture quality. Sony was the market leader in half-inch formats. Even though the Shibaden or Matsushita products were generally considered superior in performance, Sony's reliability and exemplary marketing effort, aided by a 2-3 year lead in the U.S. marketplace, more than offset its competitors' advantages.[78]

Ampex's new design under development at Los Gatos appeared to be fully competitive with the half-inch formats available in 1968, but it offered no clear advantage over them. Parity was not enough, Elkus argued: "To regain that portion of the instantaneous response market which it has lost to competition, CEPD must introduce a product with definite competitive advantages. . . ." Ampex, moreover, would have to anticipate imminent improvement in competitors' designs. "Within a year Sony, or some other competitor, may introduce devices employing automatic threading, cassette, or cartridge loading," he cautioned, features that would be highly valued in the instantaneous response segment and the playback portion of repetitive viewing markets.[79]

Elkus concluded this report by proposing that Ampex offer users a machine entirely new in design, compact, self-loading, easy to use, reliable, economical; he called it "Instacorder" (later Instavision, finally Instavideo).[80]

Elkus's analysis led to a complete reorientation of thinking about the Elk Grove Village video product line. Heretofore the driving force at Ampex had been engineering developments. Designers aimed to achieve the highest possible quality of playback signal, within cost limits, and always subject to maintainance of the near-sacred "standardization" of the recording format to enable easy interchangeability among Ampex units of any design. Now Elkus proposed that they think about (1) what the market needed; and (2) what Ampex's distinctive skills were, in order to define an appropriate niche for its products.

Having established a different way of looking at the A-V market and having gained some acceptance for his concept of the right innovative product, Elkus immediately prepared a follow-on report, issued a month later (December 12, 1968). Here he presented the "Instacorder" as the cornerstone for a renewed assault on the home video recording (HVR) market. His view was strong and clear from the first lines of the first-page summary: "Ampex belongs in the HVR market; the Instacorder should provide the foundation for a family of products which will ultimately become the standard for home video recording systems."[81] Elkus identified three segments of the consumer video business: home programming, prerecorded playback, and off-the-air recording. By home programming, he meant usage similar to home movies, which suggested product requirements similar to those he had identified for instantaneous response A-V applications. He recommended that initial efforts in the HVR market be geared toward development of the home programming segment, which he viewed at that time as the largest potential application and the one most likely to evoke initial consumer interest. While the possibility of prerecorded playback and off-the-air recording would enhance the value of an HVR system, Elkus anticipated greater obstacles to developing those uses and noted that they implied a different mix of characteristics for the equipment itself.

The December report explicitly noted that "the basic priority of features descriptive of the instantaneous response [A-V] market. . .closely match the requirements for a video recording system for the home market."[82] Elkus went on to describe the main characteristics of a design for the Instacorder to be offered in the A-V market, a design that also represented "the first step toward the development of a general line of products specifically developed for the HVR market."[83]

Elkus proposed that the Instacorder design should yield a product whose features, in order of priority, would make it:

1. "Totally reliable".
2. Easy to operate: quick to load and unload; simple to understand; rugged and adaptable to varied settings.
3. Inexpensive: $1400–$1600 initially in the A-V market for a complete monochrome system including camera; one-third that price ultimately for consumer use.
4. Compact and light-weight.
5. Modular in component design to simplify field service.
6. Simple, with automatic controls requiring few operator choices.
7. Easy to interchange tapes among models.

8. "Somewhat in excess" of the audio and video performance of standard home TV receivers.
9. Use cartridges or cassettes holding tape for 30–45 minutes and costing $10 to $12.

In accord with these priorities, Elkus urged the development of a highly compact and attractive design for an integrated camera/recorder system.

INSTAVIDEO: FROM CONCEPT TO MARKET

The engineering team at Los Gatos quickly demonstrated that Elkus's concept of a light-weight, compact recorder could be realized. Within 60 days of being given his general specifications, they had constructed a working prototype that was promptly shown to the Ampex Board of Directors. Roberts gave his blessing to an all-out effort to develop the product. Elkus, still based in California, carried forward his analysis of the market and competition, in order to develop a complete business plan for the venture. Back in Elk Grove Village, a young engineer, M. Carlos Kennedy, was assigned as project manager for Instavideo.[83]

The Ampex staff was fully occupied trying to strengthen the company's deteriorating market position in one-inch recorders. On the low-end, Ampex had to respond to growing competition from the Japanese half-inch designs. At the same time, new high-performance one-inch products from IVC (whose VTRs were distributed by the ubiquitous Bell and Howell) confronted the more expensive Ampex machines.

Roberts, while not easing any of the pressure for current profits, kept his eye on the huge consumer market for VTRs, which seemed to be coming closer. Instavideo looked to him like a vehicle to carry Ampex into the market. But just as it began to look feasible, a new competitive threat, backed by a formidable competitor, made its appearance.

In December 1968, CBS, Inc. demonstrated the latest product of Dr. Peter Goldmark's fertile imagination: "Electronic Video Recording" (EVR). Using photographic film in place of magnetic tape, EVR appeared to offer a practical alternative to the VTR for widespread use of prerecorded educational and entertainment programs. Responding to the anxiety that the EVR threat had evoked in Redwood City, Elkus produced an analysis in a report (dated April 10, 1969) entitled "Ampex vs. CBS EVR." Elkus acknowledged that EVR could become a "threat to the position of video recording in the playback portion of the repetitive viewing market," but he also identified its vulnerability. EVR, he said, "stands or falls on its ability, in conjunction with acceptable program material, to achieve a demand for wide distribution of prerecorded information."[85] EVR program costs would be highly sensitive to volume, and current A-V norms suggested most programs would fall under 300 copies. To build larger markets would take time, during which EVR and the other playback-only systems would be "all chewed up."

By May, Elkus had developed the main elements of a marketing strategy and business plan for Instavideo. Summarizing system performance specifications that had been reviewed at a management meeting on March 12, he noted that it now appeared possible to meet or exceed all of them. Moreover, a revised cost analysis indicated lower manufacturing costs than had been expected. Elkus set January, 1971 as the target for introduction of the basic monochrome version of the Instacorder, with an advanced monochrome design (stop/slow motion and editing capabilities) a year later, and a color model the following year. A cash outlay of $1.5 million would be needed to establish the business, which was projected to bring in $7 million in revenue and $750,000 in profits in fiscal 1974. The Instacorder was expected to gain 40 to 50% of a market projected to grow to 25 to 30 thousand units by 1973, three to four times its size in 1968.[86] (A sketch of the product, as it was conceived in March 1969, is reproduced in Figure 9.)

As a concept, Instavideo made sense, both in terms of the emerging video market and the needs of the Ampex Corporation. But the payoff would come only when the concept could be translated into reality—a reliable machine that could routinely be manufactured in quantity and at a low cost. To do that called for organizational capabilities that Ampex had seldom been required to demonstrate and that it proved incapable of delivering in this instance.

The performance expected of Instavideo was demanding. While skilled engineers could readily build a model that would meet specifications, a high order of product and production engineering would be required to create a design that could be mass-produced successfully. To achieve this, Ampex had to be able to apply the talents of its best engineers to development of the critical elements of the scanner and the magnetic head, and also to be able to coordinate product development with manufacturing planning.

Figure 9. Sketch of the "Instacorder"

163

Roberts, still shooting for the mass market, pushed to get a high-volume, low-cost manufacturing capability for the Instavideo venture. Elkus and others in the Instavideo team sought support for establishment of a new manufacturing facility (Taiwan was their preferred location). Roberts, pressed for cash, insisted instead on a partnership with Toamco, a Japanese joint venture formed in 1964 by Ampex and Toshiba. Toamco manufactured Ampex-designed Quad tape recorders and computer tape transports, sold in Japan by Toshiba and elsewhere by Ampex. While this move eased capital requirements, it did little to add know-how in mass-production—an art as foreign to Toamco as to Ampex—and it further complicated the task of design-manufacturing coordination, never a source of strength at Ampex under the best of circumstances.

In retrospect, it is evident that key Japanese rivals, Sony, Matsushita, and JVC, were giving high priority not only to development of designs for compact VTRs, but also to development of vital manufacturing know-how. Ampex, which had seldom competed in markets where manufacturing skill was a key factor, could not match these Japanese capabilities.

Organizational instability continued to hamper the Ampex effort to carry forward the project. In 1968, Roberts had restructured the Elk Grove Village operations, creating 4 Divisions from what had been CEPD: the Consumer Equipment Division, which manufactured and marketed home audio equipment; the Ampex Stereo Tapes Division, which produced the prerecorded United Stereo Tapes; the Ampex Service Company; and the Educational and Industrial Products Division (EIPD), which took over all the CEPD's video operations. In late 1968 Roberts assigned Ron Ballintine, who had been Ampex's resident at their joint venture with Toshiba in Japan, Toamco, as manager of the EIPD, instructing him to "turn the division around."

Narma left the Company in April, 1969 and wasn't replaced (the EIPD manager then reported to Group V.P. Robert Pappas, in Redwood City.) Ballintine soon concluded that he had been happier in the International Division, and was reassigned. Elkus succeeded him as General Manager of EIPD, finally moving his family East to Chicago.

From mid-1969 to mid-1970, a small number of engineers from Ampex and Toshiba worked on Instavideo with the objective of developing a small, lightweight, convenient, easy-to-use machine that would perform reasonably well and be, above all, reliable. EIPD lacked the engineering resources to carry its share of the joint development effort for so ambitious a product. Although Roberts had promised Elkus that he could have the best engineers in the company for Instavideo, "even if blood runs on the floor," he left Kennedy and Elkus on their own to recruit talent that was jealously guarded by other Divisions. The other managers usually proved to have more clout in a showdown, so Kennedy and Elkus, "just kids," were not able to maintain a stable engineering team. Elevated to General Manager of EIPD, Elkus still couldn't get the resources that were required for Instavideo and, in fact, had new responsibilities to distract him from it. Moreover, Roberts had decided to move the division back to California for reasons extraneous to design or manufacturing efficiency; the very move itself caused much time to be lost in the development of all the Division's products.

Instavideo's development was interrupted early in the fall of 1969 by the announcement of a new half-inch standard established by the Electronics Industry of Japan, called the EIAJ-Type 1. After discussions with Toshiba, which was working on a design of its own, and Matsushita, which was already manufacturing video recorders of this format, Ampex decided to change over for the sake of compatibility. Although Ampex and the Japanese companies could not agree on a design for cartridges built on the Type 1 format—Instavideo's priority was compactness, precluding an encased cartridge like Matsushita's—they did make interchangeability possible.

Instavideo Makes its Debut

Despite the frustrations and delays in the development of the final design for Instavideo, Elkus and Kennedy were greatly heartened by the public response to the first demonstrations of the product. Its debut in New York in September 1970 was a smash hit.[87] Even before the press conference began, Ampex had succeeded in enticing twice as many participants—some 300 reporters—as were expected to witness the unveiling of this well-kept secret.[88] And they were not disappointed. "In one bold stroke," wrote Charles Tepfer, publisher of the *ETV Newletter* of September 7, 1970:

the Ampex Corporation has shown why the reel-to-reel videotape recorder would never have made videoplayback a popular tool of the teacher, communicator, industrial supervisor, nurse, and housewife. By introducing its simple, portable (slightly larger than this newsletter and 4-1/2" thick), compact cartridge magnetic videotape system, Ampex is eliminating the hardware hangup for the user and guaranteeing that the videotape marketplace next year will be far different than it is today.

Tepfer was quoted a few days later in a letter to Elkus from an Ampex marketing executive "to the effect that . . . 'up to the introduction of Instavision it appeared that the United States industry was abdicating the cassette to Japanese industry and (he) was delighted that Ampex was taking the lead and bringing CCTV hardware back home.' "[89] Elkus recalls that one industry representative at the debut found Instavideo so attractive that he wanted to place an order for 10,000 units then and there.

Instavideo was unique for its time. It weighed less than 16 pounds, batteries included, and its monochrome camera added but 5 pounds to the user's load. (see Figure 10.) A simple stand, or "power pack," served as a battery recharger and A/C outlet.[90] Instavideo featured an automatic-loading cartridge—the most compact tape format current technology permitted; it offered slow motion, stop-action playback, and monochrome or color capability. The tape capacity was 30 minutes, or 60 in extended play; the tape format—the EIAJ-1 standard—was compatible with most other one-half-inch machines. It was priced at $1,500 with camera, $1,000 for the color recorder-player alone; $900 for a monochrome recorder-player; $800 for a monochrome player.

Especially attractive to reporters was Instavideo's versatility, for it could not only play back prerecorded cartridges—including home movies—but could also record television programs off-the-air. *Electronics,* reporting on the demonstration, believed that Ampex's gamble was that the "ability to record from TV will be the feature the public wants."[91] CBS's EVR and RCA's Selectavision, announced the previous year, offered playback only. In one of the press releases accompanying the demonstration, Elkus stated that "the forthcoming home market will best be served by equipment that not only plays back cartridge-loaded recording but permits completely portable or off-the-air recording as well. The Instavision approach permits all three in a smaller package than any other advanced to date."[92] Elkus also pointed out to *Electronics* that the price of the cartridge—$13 for 60 minutes of playing time might seem expensive, but it was "cheap compared to movie film, which costs about $80 per hour with processing."[93]

Ampex's adherence to the EIAJ-Type 1 standard also drew considerable comment. After describing Instavideo's attractive features, vid-

Figure 10. Instavideo and its 5-pound camera, 4.6-inch cartridge, 1970–1971 (Ampex archives).

eo communications consultant Ken Winslow, writing in *Educational Television*,[94] stated that

> probably the most notable development here is the adoption by the most promi-
> nent American videotape equipment manufacturer of the Type 1 standard. This
> ought to answer the concerns of those who said they would resist using video
> tape until there was an agreement on standards. This ought to help nudge the
> software distributors who have held back the release of their titles because of the
> confusion in standards. Ampex's move into Type 1 could help move us from
> parochial format philosophies and distracting comparisons . . . to the more
> vital concerns about the use of the television medium, and a unified professional
> approach—something which the application of television to education and
> training desperately needs.

Although Winslow perceived some doubt that Type 1 would be incorporated into the Japanese Industrial Standards, he argued that Ampex's
"attractive mix of convenient features" had a market: "There is tre-
mendous virtue in an agreed-upon standard for the great majority of
users who are more interested in the interchange of information (software) than they are in the world's best S/N, time base stability, and
resolution features. When left to the engineers, such things never seem
to get out of the laboratory. We need it. The time is now."[95] A
columnist in *Photographic Trade News* agreed that with Type 1, "Instavision might well become the 'genesis' standard that will make
VTR—or HVTR—truly a big business."[96]

Poniatoff added his personal plaudits. In a letter to Robert Pappas,
Group Vice President, he judged Instavideo's design and quality to be
"in accordance with the Ampex tradition of excellence."

Although Ampex intended to market Instavideo first to industrial
and educational markets, its public relations emphasized that Instavideo was "a new generation of miniature video-tape recorders and
players featuring automatic cartridge loading and designed both for
serious closed circuit television and home recording and playback
markets." Elkus called attention to the machine's convenience for
training and communications markets, and its "simplicity, reliability
and economy required for the coming home-recording and playback
market."[97]

The initial enthusiasm for Instavideo was fed by audiences at many
more demonstrations over the next year. Kennedy took the prototype
to the New York State Education convention in October 1970 and
showed it to an overflow audience who rose to their feet in applause as

the demonstration ended. The response was equally exhilarating at the National Association for Educational Broadcasters in Washington, D.C. shortly thereafter. In a joint presentation with Toshiba in Japan, and later in Europe, Instavideo evoked similar responses. Even though it still had a few bugs (usually quiescent during demonstrations), Kennedy recalled that "the audiences loved it and asked when they could buy it." Customers were so eager, in fact, that the Instavideo group had to hold back the marketing people who, it seemed, would have gladly snatched the prototypes away and sold them. Some old hands at Ampex may have found all this reminiscent of the heady days of 1956 after the VR-1000 was unveiled.

By early 1971, both Ampex and Toshiba were beginning to feel the pressure of competition, as dozens of companies jumped into the cartridge-cassette melee. The target date for putting Instavideo into production had passed and bugs still plagued the Toamco plant. In a letter to Roberts dated February 15, 1971, a top Toshiba executive analyzed the competitive situation in Japan and determined that: (a) Instavideo's round cartridge should be intensively promoted as a standard for the industry; (b) Instavideo should be marketed as soon as possible, preferably by June; and (c) Instavideo's pricing should be reconsidered in light of Matsushita's recently introduced, low-price half-inch machine. Toshiba was not yet worried about the impact of the 3/4-inch cassette recorders introduced by Sony and others, in part because it expected Sony to manufacture an Instavideo-type machine as well. It was the latter expectation that convinced Toshiba that Instavideo must reach the market soon if it was to succeed.

Meanwhile, Elkus adapted his speeches to the emerging situation, expanding his analysis of the benefits of Instavideo's unique abilities, underplaying the importance of the EIAJ standard, and stressing that the home market was still distant. His vision of programming for cartridge television embraced five categories: constructive leisure, classic works of art, emotional stimulants, background information, and pure entertainment. All but the latter were "personally involving," and could be best developed when users were intimately involved with the programming device and the program itself. Hence the public needed to be educated to the value of visual media, of the benefits of involvement in this "revolutionary new way of communicating with one another." At the same time, he stressed that the home market would take several years to develop, that "major progress may not be made until 1975."

Elkus increasingly emphasized that the Instavideo system was designed to sell itself regardless "of the availability or appropriateness of existing software." Since a universal standard did not appear to be forthcoming and since Ampex certainly was not in a position to provide software, Instavideo had to be sold on its own merits as a self-contained device. In written materials on Instavideo as well, he elaborated on its versatility for training in industry, business, law enforcement, medicine, and education, where a portable instrument capable of instant playback could provide benefits not offered by any other system. He also explained that Instavideo was part of a whole line of EIPD products, and renamed the division's 1-inch machines to reflect their new position in that line as Video Production Recorders (VPRs), higher performance units suitable for high quality production.[98]

Instavideo shared a full stage with many other competitors in Cannes, France, at the First International Cartridge TV, Videocassette and Video Disc Conference (VIDCA) in April 1971. Carlos Kennedy represented Ampex, reading a speech prepared by Elkus that emphasized the necessity of educating the public to the "benefits of involvement in CTV" so that they will "begin to accept the video cartridge system and its inherent versatility as a normal part of every day living." According to *Billboard* of May 15, 1971, he "predicted that pure entertainment programming, unless it can be offered at a very low cost, will probably not stand as a principal reason for the sale of video cartridge systems." Such a prediction confronted the playback-only systems at their core.

In May 1971, Elkus issued one more report on Instavideo. In it he updated his discussion of the competition and evaluated Instavideo's progress thus far. He again stressed the need to focus all efforts on what Instavideo could do best as the smallest, most practical machine for the instantaneous response market. Although advertising rhetoric claimed that Instavideo could do everything, Elkus modified his earlier analysis with the conclusion that a machine competitive in the prerecorded playback market would have to be designed with different criteria, different priorities in mind than was the current Instavideo machine. The development of a machine for prerecorded playback, he suggested, would require $700,000 to $1 million in further development funds. Meanwhile, he argued, Ampex should pour its efforts into getting the current model debugged and in the marketplace in order to maintain its lead. Elkus stressed that the focused strategy he had recommended for Instavideo had a good chance of succeeding because of the product's inherent strengths. Unlike Roberts, who continued to think in terms of a mass market for Instavideo, Elkus did not want to get distracted from his current priority: getting Instavideo into production. Development of a design more appropriate for prerecorded playback could be carried on later, after the market had given some indication of what requirements such a design should embody.[99]

But Instavideo could not be sold until the Toamco plant was ready to manufacture it. Design engineers had to get the bugs out of the product, tools had to be built and tested and the manufacturing process organized. Most of the crucial tools were made in the U.S. and flown to Japan; their designer, Jack Riccardo, spent several months on location helping Toamco production engineers set them up. But both Ampex and Toshiba encountered difficulties inherent in a production operation so far removed geographically from the design lab. That it was an international effort exacerbated the communication problems. Because, as Elkus later pointed out, Ampex had given Instavideo to Toamco "before we should have," engineers in California were continuing to make design changes while those in Japan were trying to make the machine manufacturable.[100]

Neither the Ampex engineers assigned to Instavideo nor the engineers from Toshiba had much experience in making a complicated but relatively low-cost, high-volume product like Instavideo truly manufacturable. Elkus lamented this lack of experience, especially in comparison to their Japanese competitors, in his May 1971 report. He suggested that an outside firm do a "value engineering analysis" of Instavideo in order to help identify possible improvements, in particular those resulting in lower manufacturing costs within the given product specifications. A tone of urgency pervaded the report, to no avail. As Elkus later exclaimed: "We didn't know what the hell they were doing and they didn't know what the hell we were doing and we didn't know how to speak to them about our problems." Most important, however, were company-wide developments that threatened Ampex—and even its most promising innovations—with oblivion.

FINANCIAL CRISIS

In 1962, at the dawn of the age of the "glamour stock" on Wall Street, Ampex had been regarded as one of a select group of young, technology-based growth companies, along with firms like Texas Instruments, Polaroid, and Hewlett-Packard. Ampex grew throughout the sixties at a pace comparable to its glamorous brethren.[101] New plants were built or old ones expanded in Colorado, Illinois, Alabama and California; overseas operations grew even more rapidly. The number of employees nearly tripled, from under 5,000 to over 14,000. New products proliferated in all areas, and Ampex could continually claim that 60%–80% of its revenues were from products no more than three years old. The record of growth seemed to confirm the success of Roberts and his management methods. At the end of the decade the opportunities inherent in the core businesses, and particularly in video and information systems products, seemed even greater than they had in 1962. In 1969 Ampex stock continued to trade at prices as high as 35 times earnings. But behind the facade of accounting results, the foundations of the business grew increasingly less secure.

Although it appeared to be reporting improved earnings each year under Roberts' leadership, a closer analysis reveals that its profitability was steadily deteriorating. Figure 11 shows the main trends in financial performance during the first nine years after Roberts took office. Ampex's revenues grew at an average rate of 18% annually during the 1960s, and net profits grew apace. But assets expanded much more rapidly, with heavy capital obligations in connection with new products like Videofile and ambitious plant expansions like those in Elk Grove and Colorado Springs. Invested capital grew at a compound rate of 26% per year and return on investment, never very high at Ampex, declined steadily.

Roberts's proclaimed intentions to keep Ampex on the track of rapid growth in sales and profits—at least 15% per year—translated into unceasing pressure on subordinates to show quarter-to-quarter improvements in net profits.[102] But the investments required to maintain the illusion of rapid growth only diluted the company's profitability, which was insufficient to pay the costs of growth. Hence that burden was carried for the most part by an expansion of long-term debt, which grew ten-fold to reach $150 million in 1969.[103]

Ampex was masquerading as a successful high-growth company in the period of Wall Street's rage for "glamour," but it could not earn the profits necessary to fuel its growth.[104] Figure 12 compares Ampex's financial performance with that of two other companies that experienced comparable rates of sales growth and technological change in the 1960s. While the capital employed by Ampex was increasing more rapidly, its return on capital was only a fraction of what

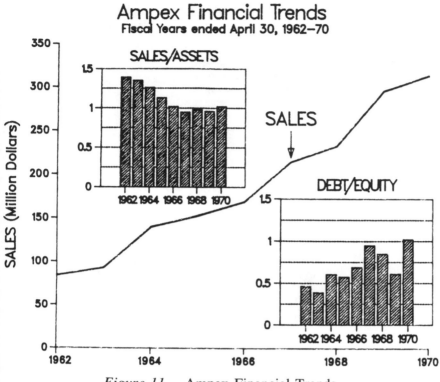

Figure 11. Ampex Financial Trends

the others were earning. The comparison with neighboring Hewlett-Packard is especially revealing. From 1961 through 1969, these two young firms grew at the same pace, with revenues more or less equal throughout. However, Hewlett-Packard's return on invested capital was double that of Ampex in 1961 (Ampex's fiscal 1962) and more than three times Ampex's by 1969.

Low returns meant that Ampex chronically lacked sufficient funds to support the many investment opportunities presented by a rapidly changing technology and broadening markets. To make matters worse, some of its acquisitions were a drain, rather than a source of profits. Some ventures closer to the core, while initially promising, proved to have no sustainable competitive advantage. Consumer audio, for example, slid into the red as the market changed and Ampex was unable to adapt.

Ampex had reported a dip in profits in 1968, but it was not clear until mid-1971 that this was but the tip of the iceberg. At the end of

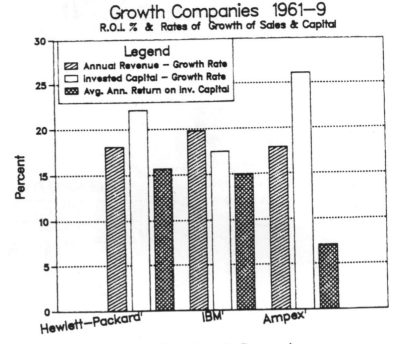

Figure 12.   Growth Companies

fiscal 1971, the company reported a loss of $12 million. When Roberts resigned as president and chief executive officer, Arthur Hausman, an engineer who had joined the company in 1960 as vice president and director of Research, was picked to succeed him. Roberts stayed on as chairman for a few months, then was relieved by Richard Elkus, Sr., a member of the board since 1959. The next January, the company announced that it expected a loss of $40 million for the year, a sum that had mounted to nearly $90 million by the time the auditors released the accounts for fiscal 1972.[105]

Announcing their intention to focus the company on what it could do best—make and market professional audio and video equipment—President Hausman and Chairman Elkus took drastic measures to keep Ampex afloat. They recentralized the company, immediately reducing its divisions from 12 to 7; later they would consolidate into 3.[106] Over the next two years they sold several subsidiaries and closed the Elk Grove plant, discontinuing the production of consumer audio recorders (the video operation had already moved to California). They hesitated to terminate the Instavideo project, continuing to suggest, at least in public, that it was on track. *Electronic News* of March 1972 quoted Charles A. Steinberg, vice president for Audio/Visual, as expecting production of Instavideo by the end of the year.[107] But the company's resources no longer could support so ambitious a venture. Recognizing that significant further investments would have to be made to perfect the design, get it into full production, and launch it on the market, management apparently believed it had no choice but to withdraw from this once highly promising project. In October 1972, the senior Elkus announced the decision to drop the development of Instavideo altogether.[108] Henceforth Ampex's only role in home video would be as a supplier of blank tape for the products of former competitors.[109]

Thus Ampex's almost-entry in the home video contest was snuffed out before it could prove itself in the marketplace.[110] Whether or not Ampex could have modified Instavideo for home use and adapted it to further advances in technology; whether Toamco could have produced efficiently at high volume; whether it could have worked out a profitable marketing arrangement with a successful consumer company—all these questions are moot. Instavideo simply did not get that far.

With the demise of Instavideo, American industry lost its best chance to participate in the mass market for video cassette recorders. During the next decade, when the VCR emerged as one of the preeminent growth businesses of the 1980s, every new design was developed and manufactured in Europe or Japan.[111] The lessons inherent in Ampex's failure have, therefore, a broader significance to American management.

The essence of this story lies in the actions of corporate management, rather than in the management of the Instavideo project itself.[112] The way in which William Roberts and his associates pursued "dynamic growth" created the forces that ultimately denied Ampex its greatest growth opportunity. Roberts failed to comprehend that the remarkable growth of Ampex in its first decade was rooted in its focus on a few businesses in which technological leadership could be translated into defensible competitive advantage. Driven by Roberts's emphasis on diversity and quarterly increases in reported earnings, Ampex was committed to too many markets and unable to establish and defend profitable long-term advantage in the most promising among them.

Instavideo proceeded far enough to demonstrate Ampex's command of leading-edge technology. Subsequent events have confirmed the soundness of the project leaders' comprehension of the market opportunity. Ampex lost the chance to build on those strengths because of top management's inability to focus the firm's human and financial resources productively on the opportunity before it.

APPENDIX A: EARLY HELICAL SCANNER INVENTIONS AT AMPEX

Alex Maxey set out in 1955 to simplify VTR design (compared to the complex Quad) by the use of one head in a format that would fit an entire video field onto a single "track." He undertook to replace the 4-head, multiple track design then nearing completion at Ampex with a simpler approach that still retained the core idea of a swiftly rotating head sweeping by a (relatively) slowly moving tape.

Late in 1955 Maxey made his first machine with a single rotating head. Built "for practice," the device was called the "one-headed hoo-hah" among the engineers. In this experiment Maxey curled the tape lengthwise some 370 degrees to form a tube around a rotating scanning drum; a single head on the drum wrote transversely in a spiral on the inside of the tube. This design was so crude, according to Maxey, that it was a question whether the tracks on the tape were recorded or "engraved." Although this machine could record and play, the tubular wrap was so awkward that one could not envision a practical tape transport built on this concept.[113]

In Maxey's first one-head helical prototype, dubbed "Pan-I," the two-inch tape was wrapped helically in an omega pattern, covering just under 360 degrees of the circumference of a vertical mandrel. In this version, the tape was held in place by a guide similar to the one used on the VR-1000. Since the writing speed was the same as the Quad's, 1570 inches per second, the mandrel had to be very large in diameter—eight and one-quarter inches—in order to get a track long enough to record a full field. The first prototype of this scanner was completed in late 1956, a patent filed for in March of 1958 (see Figure 4A).[114]

The next step toward simplification was to develop a system to regulate tension and tape movement in order to eliminate the guide mechanism. (see Figure 4B.) This technology was incorporated into several two-inch, one-headed prototypes in the late 1950s, including the so-called "NAB machines" and the VR-8000, Ampex's first publicly announced helical recorder. (see Figure 4C.)

In parallel with the efforts that led to Pan-I and the VR-8000, Maxey was involved in several other projects. As early as 1957 he began working on elements of a "simple-minded machine for a simple-minded market," a model he called "Junior." It used a two-inch tape, which was wrapped helically around a drum with no guide. It was conceived as a two-headed machine with a 180 degree wrap. A writing speed of 641 inches per second, a tape speed of five inches per second, and a narrower track meant a significant savings over the Quad in the quantity of recording medium used.[115]

During this period Maxey was also working on a video recorder for NASA, particularly challenging because it had to be very compact yet perform extremely well. He had experimented with a compact tape-threading mechanism in a model he called "Ham-pan" early in 1958; this machine had used a female guide to hold the tape in a wrap of somewhat over 180 degrees around a small, Quad-size drum, but it was still too cumbersome for use in space. In a subsequent machine built early in 1961, Maxey stacked the two reels, thus solving a number of space and stability problems at once. The final machine weighed only 35 pounds and occupied less than one cubic foot.[116]

ACKNOWLEDGMENTS

The research on which this paper is based included interviews with most of the principals in the events described and examination of many documents, some of the most important of which were provided from the private files of the interviewees. To say that without the generous cooperation of these many people this study could not have been completed is to understate our obligation to them.

Those who helped us include three of the five members of the Ampex team that first invented the VTR, all four General Managers of the CEPD/EIPD operations, and nearly a dozen other engineers and managers who had a direct role in the events described. We hope that all will be able to find some satisfaction in having the story told. We want to offer a special word of thanks to Charles Ginsburg, who welcomed our initial inquiries and patiently assisted us throughout. Without his leadership 30 years ago the "video age" might have been a great deal later in arriving.

NOTES AND REFERENCES

1. "The Climate For Innovation In Industry," by Richard S. Rosenbloom and William J. Abernathy, *Research Policy,* 11: 4 (1982), 209–225.

2. On Poniatoff, see Norman Eisenberg, "High Fidelity Pathfinders: The Men Who Made an Industry. Alexander Poniatoff," *High Fidelity Magazine* (September 1977), 72–73; "Alexander Poniatoff and His Unpretentious Approach to Life,"

Broadcasting (December 15, 1969), 89. Poniatoff was born in Russia and studied mechanical engineering first there, then at Karlsruhe in Germany. After serving as a pilot for the tsar at the beginning of World War I, he found himself on the wrong side of the civil war that followed the Revolution of 1917. He fled to China, where he worked as an engineer for several years.

3.   There is a striking similarity in the origins of Sony Corporation, later to become Ampex's most formidable rival. During the war, Masaru Ibuka, Sony's founder, operated a small firm making precision equipment for the Japanese Navy. Afterward, Ibuka struggled to keep his tiny firm alive. Now joined by Akio Morita, he searched for a product that could be sold in volume. A member of the American Occupation forces showed him an early tape recorder and he "was immediately convinced that this was the product they had been looking for." And so it was.

For more of this story and the origins and growth of Sony, see *Sony Vision* by Nick Lyons, Crown Publishers, 1976.

4.   On the development of the Ampex audio recorder, see Harold W. Lindsay, "Magnetic Recording, Part I," *The Sound Engineering Magazine* (December 1977), 38–44, and "Magnetic Recording, Part I," *ibid.* (January 1978), 40–44; and John T. Mullin, "Creating the Craft of Tape Recording," *High Fidelity Magazine* (April 1976), 62–67. Poniatoff had been considering broadcast turntables for Ampex's new business and hired Lindsay as a consultant to help explore the opportunities. Lindsay recalled that Poniatoff lost no time in responding to the challenge of developing a tape recorder instead (see Lindsay, part I, 39).

5.   After a later public offering Poniatoff owned less than 1% of the equity, while the partners continued in control with some 20%.

6.   Sarnoff's speech was covered in the *Wall Street Journal* (September 28, 1951) and the *New York Times* (September 28, 1951), 39. Broadcasting (October 1, 1951), 27. quoted the speech extensively. Two months later, *Business Week* (December 1, 1951), 112. covered an announcement by Crosby Enterprises of an experimental video recorder that would be ready to record "standard video pictures" in six months. RCA's machine was first demonstrated in late 1953; see: *Business Week* (December 5, 1953), 34; *Electronics* 27 (January 1954), 5–6. A technical description is in Albert Abramson, "A Short History of Television Recording," *Journal of the SMPTE* 64 (Feb. 1955), 75.

7.   Eisenberg, 72.

8.   An Italian, Luigi Marzocchi, had been working on methods of rotary head recording, in the early 1930s. He obtained patents for a variety of conceptual designs, but a great deal remained to be done to incorporate them in a practical machine.

Marzocchi's Italian patent was applied for in July of 1936, issued in April of 1937; his British patent, entitled "System of Electromagnetic Sound Recording on Metallic Plates, Ribbons, or Wires," was applied for in June of 1937, granted (No. 497,800) on December 28, 1938; a United States patent (2,245,286) was issued on June 10, 1941.

9.   The account that follows is based on interviews with Charles Ginsburg, Fred Pfost, and Alex Maxey in 1981 and 1982; a contemporary account written by Ginsburg: "The Ampex Videotape Recorder: An Evolution," a talk given on October 5, 1957, at the 82nd Convention of the Society for Motion Picture and Television

Engineers (SMPTE) in Philadelphia, (Ginsburg repeated the talk in part at the 122nd convention of the SMTPE in New York City on November 13, 1980); and on Ginsburg's Schoenberg Memorial Lecture to the Royal Television Society, London, November 5, 1981 published in *Television,* The Journal of The Royal Television Society, 18: (12), 11. In 1981, Ginsburg was vice president for Advanced Technology Planning at Ampex.

10.   Dolby left Ampex when the project ended, completed his education, and then turned his talents to other inventions in the field of magnetic recording. The noise reduction system he developed for stereo music has made his name a household word.

11.   "Quadruplex," shortened to "Quad," became the generic term for videorecorders built on the VR-1000 format.

12.   "Market Survey for VTR Equipment in TV Stations," December 12, 1955. From R. S. Isberg to P. L. Gundy.

13.   Ginsburg in the preface to his talk at the SMPTE in November, 1980.

14.   See: Ginsburg, "The Ampex Videotape Recorder," 9–10.

Variations in temperature and humidity, as well as wear on the heads, contributed to the complexity of this technical problem. Maxey found several ways of addressing it. The penetration of the heads into the tape was determined by the grooved guide that held the tape against the rotating head drum. The effects of the variables could be influenced, Maxey determined, by controlling the position of the guide (the method soon adopted), and also by changing the vacuum in grooves added adjacent to the head scanning region, or by modifying the longitudinal tape tension.

15.   *Ibid.,* 10.

16.   *Ibid.,* 10. A demonstration using the team's other operable VTR was given simultaneously for the press in Redwood City. See *Wall Street Journal* (April 16, 1956), 16. *New York Times,* (April 22, 1956), II, 13. A published paper followed in May: Charles P. Ginsburg, "A New Magnetic Video Recording System," *Journal of the SMPTE* 65 (May 1956), 302–304. More extensive technical discussions were published a year later by members of the team: Ginsburg, "Comprehensive Description of the Ampex Video Tape Recorder," *Journal of the SMPTE* 66 (April 1957), 177–182; Charles E. Anderson, "The Modulation System of the Ampex Video Tape Recorder," in ibid., 182–184; Ray M. Dolby, "Rotary-Head Switching in the Ampex Video Tape Recorder," *ibid.,* 184–188. Over the next few years Ginsburg, as head of the team, received several awards and prizes for this revolutionary innovation.

17.   The recollections of Ginsburg and Alex Maxey in conversation with Rosenbloom on October 7, 1980. Gundy was manager of the Audio Division in April 1956.

18.   Anecdote related by William C. Connally of CBS, when he introduced Ginsburg at the SMPTE panel in November 1980.

19.   *Journal of the SMPTE* (May, 1957), 257.

20.   RCA's share grew to about 35% and remained there throughout the mid-1960s and early 1970s; Ampex held about two-thirds of the Quad market during that period.

21.   Ampex's international business grew rapidly, representing 12% of company sales in fiscal 1960, 16% in 1961, 33% in 1970.

22.   Poniatoff thereafter played no major role in Ampex's day-to-day operations, but he remained chairman of the board until 1970 and was the living symbol of the company until his death in 1980.

23. Ampex had sold expensive audio tape recorders for home use to wealthy afficionados as early as 1954, but the first big step toward a mass market for tape recorders was the invention of four-track stereo tape.

To assure a supply of quality tape for all its recording divisions, Ampex—which had been working with 3M on magnetic tape since 1947—invested in a small tape manufacturer in 1957, merging with it through an exchange of stock in 1959. The Orr Industries plant, in Opelika, Alabama, became the Magnetic Tape Division in 1960.

24. See Robert Lubar, "Five Little Ampexes and How They Grew," *Fortune* (April 1960), 116+.

25. See "Comeback for Ampex," *Time* (December 21, 1962), 69; *Wall Street Journal* (September 17, 1962), 1+.

26. Ampex never published information about profitability of individual product lines. Several former executives, in interviews, commented on the extraordinary margins earned on the Quad over the years. One suggested that during the 1960s "broadcast equipment profits were 120% of the net for the company as a whole."

27. In 1962 the company demonstrated its first solid-state unit, the VR-1100. The Editec electronic editor was available in 1963, and the first direct color recorder, the VR-2000, in 1964. A compact version of the VR-2000, the VR-1200, appeared two years later. In 1967 Ampex delighted sports fans with the HS-100, a slow-motion, instant-replay disc recorder. Ampex's first cassette-loading Quad—the ACR-25, a machine designed expressly for commercials—came out in 1970.

28. The proportion of Ampex's business devoted to military products peaked in 1963 at 42%.

29. *Ibid.*, and "William E. Roberts. In the California Air." *Forbes* (September 1, 1966), 22.

30. For a description of Videofile, see *Magazine of Wall Street* (September 30, 1967), 18–19; Annual Reports of 1967, 1968.

31. "Ampex Reaping Fruits of Research," Financial World (December 8, 1965), 13.

32. *Wall Street Journal* (April 16, 1956), 16.

This view conflicted with the pessimistic conclusions of Ampex's market survey, which said that because of the high cost of the equipment, the market for VTR in the home "does not seem to be promising in the forseeable future."

33. *New York Times* (April 22, 1956), II., 13.

34. Carl M. Loeb, Rhoades & Co., "Ampex Corporation: A Study of the Leading Manufacturer of Professional Tape Recording Machinery," (July 1958), 8.

35. The so-called "helical" scanner permitted use of a simpler design than the quad while retaining the idea of a swiftly rotating head sweeping by a (relatively) slowly moving tape. The combination of motions of tape and scanner produced a straight track slanted at an acute angle across the tape, a track long enough to incorporate signals for an entire field. This format became known as "helical scan," because to achieve it, the tape in early designs was wrapped around the scanning drum in a helical pattern.

In 1950, Earl Masterson of RCA Laboratories filed a patent application for a "method of magnetic recording of high frequency signals" employing a rotating head to scan a broad magnetic tape wrapped helically around a cylindrical scanner. The patent (2,773,120) was issued December 4, 1956. There is no evidence that this design was ever used in a working recorder. During the mid-1950s, engineers at Telefunken in Germany and Toshiba in Japan also experimented with rotating-head scanners using a helical tape path. While Toshiba carried its effort through to a working model, that stage was not reached until after its engineers had "borrowed" elements of the successful Ampex broadcast recorder.

36. As it turned out, machines using the VR-1000's quadruplex format would dominate in broadcasting for 20 years before machines using the helical scanner could compete with them in performance.

37. Three months later, however, at the SMPTE convention, Toshiba of Japan won the attention that Ampex—and perhaps others—might have had by presenting the first public paper on helical recording, much to the surprise of the majority of those present.

The paper was read at the SMPTE of May 5, 1960, and published the following December: Norikazu Sawazaki, et al., "A New Video Tape Recording System," *Journal of the SMPTE* (December 1960), 868–871.

38. The patent for this design was no. 3,377,436, issued April 9, 1968. The VR-1500 was priced at $12,000. A modified design, called the VR-660, was offered to broadcasters, beginning in 1964.

39. This relatively simple though effective lubrication method not only eased capstan load, tape wear, and timebase errors in general, but it also tended to minimize differential tape stretch—an enemy of interchangeability—around the scanning assembly.

This key patent was issued as no. 3,404,241 on October 1, 1968, with the title "Helical Scan Magnetic Tape Apparatus with Self-Energized Air Lubrication."

40. Maxey returned to Ampex in September 1977; he is now senior staff engineer in the Data Systems Division.

41. When Roberts broke up the "five little Ampexes," merging consumer audio with the professional business, Narma survived as head of engineering.

42. Ellmore, a former physics professor, had been chief engineer of Ampex Audio, Inc. When Roberts consolidated the organization he remained to work under Narma. Siebert was a skilled tool and die maker.

43. See "Ampex Reaping Fruits of Research"; "Ampex Makes a Play for the Home Market," *Business Week* (June 17, 1967), 165–174.

44. See "Marketing Strategies for Tape Recorders," *Merchandising Week* (May 3, 1965), 39–40+.

Roberts claimed in an interview in 1965 that Ampex was "entrenched in the mid-price range" of the audio market and also had "the Cadillac" of home audio equipment. (see "Ampex Reaping Fruits of Research.") By Mid-1956, Ampex's pre-recorded tape business, United Stereo Tapes, offered 1,600 selections from 44 record companies.

45. Chris H. Schaeffer, et al., *Video Tape Recording: New Markets and Products* (1965), 16. Cinerama, a U.S. company, had just bought a controlling share, 51%, of Telcan.

46. The three were Delmar Johnson, William Boylan, and James Quinn. Maxey had already left Ampex and Streets did the honors.

47. Ampex had offered a VR-1500 in a luxury "home entertainment console," called the "Signature V," as a Neiman-Marcus Christmas special in 1963. At

$30,000, it hardly qualified as consumer video, but one unit was sold. As the 150-page owner's manual illustrates, the VR-1500 was a very complicated machine for a non-technical person to operate.

48. By November 1963 only 300 had been delivered worldwide.

49. They were separated from the staff of the fast-growing audio operation, going to work at first in a small leased facility, moving in November to a new plant, about a mile from the audio operation.

50. A machine with one head had to have the tape wrapped around all 360 degrees of the scanner's circumference. The resulting friction between scanner and tape made this impractical for routine use. But the invention of the "air-bearing" two-section scanner opened the door to single-head designs.

51. In fact, the design yielded a machine of far higher performance than either they or Redwood City had intended should come from Chicago.

52. In a console with a built-in 9" monitor, the Sony machine (called the TCV 2010) was priced at $995.

53. "Tape-It-Yourself TV," *Life* (September 17, 1965), 56–60. *Life* referred to Sony and Ampex as "giant companies" although their combined revenues that year were about one-eighth those of their rival, RCA.

54. *Merchandising Week* (May 9, 1966), 13.

55. "Ampex Introduces Line of Home TV Recorders," *Wall Street Journal* (June 24, 1965), 4.

56. "Ampex Reaping Fruits of Research."

57. Trux to Rosenbloom and Freeze, October 22, 1981; Latter to Rosenbloom, February 13, 1981. Trux took a VR-7000 prototype to Hawaii since what would be called the VR-6275 was not yet ready.

58. "Volume Up for TV Recording," *Business Week* (March 12, 1966).

59. The Bell and Howell alumni at Elk Grove Village were quick to characterize their new target as the "16 mm" market, referring to the motion picture film used universally for "audio-visual" applications.

60. In mid-1967 the Annual Report disclosed that more than 800 people had completed its five-day course during its first year of operation.

61. Trux to Rosenbloom and Freeze, October 22, 1981.

62. Most of the field problems seemed to pertain to the scanner assembly. Some later suggested that problems often were the result of poor quality recording tape, which left oxide on the rotating drum or clogged the head itself.

63. The market for conventional audio-visual equipment (mainly 16 mm. film projectors) was itself quite small. In 1965, total sales of 16 mm projectors in the U.S. were estimated at 45,000 units worth $42 million. See: "Market Review: Non-Theatrical Film and Audio-Visual—1965," by Thomas W. Hope, *Journal of the SMPTE*, 75: (December 1966), 1204.

64. IVC was a new firm in Mountain View, California, started by one of the founders of Memorex and several Ampex alumni.

65. Still focused on consumers, they first showed this model at the Chicago Music Show.

66. K. Iwama, president of Sony, to R. Rosenbloom, in Tokyo, July 17, 1980.

67. A subsequent market study (see Elkus Report, discussed below) reported that

IBM found that its eight Sony machines required only one service call over an eight-month period. Other users reported similar experiences.

68.   The color recorder was priced at $4995; Ampex guaranteed the adaptability of their other models to color at conversion costs ranging from $500 to $1000. The first designs with color circuitry required the use of a modified TV receiver for playback.

69.   By 1970, Elk Grove Village was marketing a broadcast quality production machine, the VR-7900 (later called the VPR-7900), father of the one-inch helicals that would start replacing Redwood City's Quads in the late 1970s.

70.   Quoted by Elkus in his first Report (see below).

71.   After undergraduate studies in economics and political science, Elkus had earned an MBA in 1959. He met Roberts when his own company began to sell to Ampex. He says that he was intrigued by the opportunity to get involved in strategic planning, and further attracted by the promise that he would soon be given a division to manage.

72.   Elkus to Rosenbloom, September 17, 1981, and to Freeze, October 19, 1981.

73.   The first two reports—later labeled "sections" of a report called "Introducing the Instacorder," November 12, 1968–May 22, 1969—were entitled: I. Introducing the Instacorder—"The Audio Visual Scratch Pad" (November 12, 1968). II. Introducing the Instacorder as a Family of Products (December 12, 1968). Later they were augmented by three additional installments: III. Ampex vs. CBS EVR (April 10, 1969) IV. Introducing the Instacorder as a Family of Products, Part II (May 1, 1969). V. Introducing the Instacorder as a Family of Products: Preliminary Marketing Strategy (May 22, 1969).

74.   It is interesting that Elkus's very thorough and sophisticated analysis of the market dealt solely with U.S. customers, even though he clearly contemplated the possibility of production in Japan. In fiscal 1968, 29% of Ampex's sales came from outside the U.S., and the company had five manufacturing plants abroad.

75.   Elkus report, I, 1.

76.   *Ibid.*, 40.

77.   Further development of a smaller, lighter one-inch machine was recommended by Elkus only as the last of several, more promising alternatives. See Part IV of "Introducing the Instacorder," Section I.

78.   Elkus report, I, 5.

79.   Elkus, I., 37.

80.   To Elkus, the phrase "audio-visual scratch pad" evoked an image of a very convenient instant-playback machine that one could use as easily as a cassette audio recorder—as easily as a note pad. The name adopted in March of 1971 was Instavideo—sometimes Insta-Video—because the other two names turned out to be trademarks of other companies. "Instavideo" is frequently used in this paper even when referring to the period prior to March, 1971.

81.   Elkus, II, i.

82.   Elkus, II, 28.

83.   *Ibid.*

84.   Kennedy was an electrical engineer who had been in charge of Elk Grove's video cameras and other VTR accessories since the spring of 1968; he continued to work on cameras for a few months after he joined the Instavideo project.

85. Elkus III, 26.

86. Elkus IV, 28.

87. See "Portable VTR Reproduces Color," *Electronics* (September 14, 1970), 155–158.

88. One reporter noted that one of his colleagues had his press kit "filched—a most unusually occurance" (*Investor's Reader,* October 7, 1970.)

89. L. A. Wortman to Elkus, September 10, 1970.

90. The stand also contained the optional color circuitry, since it was pointless for the color circuitry to be included in the recorder itself until a color camera was available.

91. *Electronics* (September 14, 1970).

92. Ampex press release, September 2, 1970.

93. *Electronics* (September 14, 1970).

94. *Educational Television* (October, 1970).

95. *Ibid.*

96. *Photographic Trade News* (October 1, 1970).

97. Press release, September 2, 1970.

98. When the Ampex one-inch machines began to be produced in California, their market focus was shifted from all-purpose closed-circuit machines to production—high quality—machines; hence they were rechristened "Video Production Recorders—VPRs." Until Ampex and Sony, in cooperation with other companies, compromised on a standard format for one-inch machines, called "Type C," in the mid--1970s, these Ampex machines used a tape format essentially the same as that of the Elk Grove machines built in the mid- and late-1960s.

99. Richard J. Elkus, Jr., "Instavideo Product Plan. Preliminary Analysis, May 18, 1971."

100. Ricardo believes that these changes compromised Ampex's ability to get into production as early as it should have.

101. From 1961 through 1969, Ampex's revenues grew at a compound rate of 17.9% annually, compared to 17.2%, 20.6%, and 18.1%, respectively, for the other three companies over the same period.

102. *Business Week* (June 17, 1967), 65–71. Latter and Narma, running the profitless Elk Grove operation, felt particularly victimized by this attitude from management; they believed in their products and wanted more support in these early stages. (From conversations much later, in 1981, with Rosenbloom.)

103. One increase in equity was made in November of 1968 by forcing conversion of $30 million of 5 1/4% bonds to common stock.

104. See the early warnings—atypical of the business press on Ampex—on Ampex's lagging profits in *Forbes* (September 1, 1966), 22; and (November 15, 1968), 82.

105. The company's stock plunged to $11 (and would continue to drop until 1974 to $2) and Roberts became the target of a class action suit in which an irate stockholder charged him with selling out his Ampex shares at $45 in 1969 on the basis of inside information about the company's true financial condition. According to the *New York Times* (April 26, 1979), D1, the suit against Ampex and Roberts was settled out of court for $7.8 million in 1976; Roberts died of a heart attack in 1977.

106. "A Painful Attempt to Aid Ampex," *Business Week* (February 12, 1972), 17.

107. "Recording Lines Played Lead in Ampex Drama," p. 20. Ampex's videotape products manager was quoted in April as saying that "the firm has no plans to compete in the consumer CTV [cartridge television] hardware or prerecorded software markets."

"Ampex to Limit its Consumer CTV Role," *Merchandising Week* (April 24, 1972), 8.

108. "Ampex Axes Instavideo," *Electronic News* (October 16, 1972), 48.

109. Ampex tried to sell its part in the Instavideo venture to Magnavox, which was looking for an opportunity in the home video arena. The deal fell through, however, as Philips acquired Magnavox and Toshiba lost interest as well.

110. In 1973 Instavideo was allowed one last chance. In Washington, D.C., some members of ABC's staff remembered its unique, attractive compactness—and hence portability—in contrast to the heavy, bulky U-Matic; they bought two or three of the prototype models for use in what would later be called "elective news gathering." Fitted with a modified time-base corrector, these Instavideos were even used experimentally for on-the-air broadcasting. In the end, however, the U-Matic's performance and lower cost—for now U-Matics were in high-volume production—persuaded ABC to drop the Instavideos.

111. A rival American machine was brought to market in late 1972, just as Instavideo was going under, by a small firm called Cartridge Television Inc. (CTI). Their product was significantly inferior to Instavideo in performance and was a failure in the market. CTI was bankrupt within a year.

RCA tried to develop a VCR for consumers after years of neglect of the technology. Recognizing the advantages of the VHS and Betamax designs, RCA soon abandoned its own developments and captured the largest share of the U.S. market as a distributor for Japanese-made VHS machines.

112. Technical or marketing failure in a given innovation need not have terminated the efforts of a company in this field. As noted at the beginning of this paper, the most successful Japanese innovators were persistent pioneers who experienced significant failures before achieving a major success. Japan Victor (JVC), which developed the most successful design, the VHS, was approximately the same size as Ampex in 1972, lacked substantial marketing resources outside of Japan, and had suffered setbacks in a series of previous efforts at video innovation. Yet, as a consequence of the VHS innovation, JVC's worldwide sales of VCRs exceeded $2 billion by 1982.

113. The patent on this design was no. 2,912,518, issued November 10, 1959.

114. Patent no. 2,998,495, issued August 29, 1961.

115. The patent for this design was no. 3,377,436, issued April 9, 1968.

116. The patents for the NASA design were No. 3,159,501, issued December 1, 1964 and No. 3,189,289, issued June 15, 1965.

RELATING TECHNOLOGICAL CHANGE AND LEARNING BY DOING

John M. Dutton and Annie Thomas

I. INTRODUCTION

The business sector provides fertile ground for inventing and developing new technologies. Business firms are a major target for efforts to discover new technologies and to incorporate them into their host societies. But this assignment is a new one for business firms. Only within the last century have firms like AT&T, DuPont, General Electric, General Motors, and Westinghouse explicitly incorporated into their policies provisions for allocating and organizing resources to the goal of systematically generating new technology.

In business firms technologies are especially visible: (1) in firms'

Research on Technological Innovation, Management and Policy
Volume 2, pages 187–224
Copyright © 1985 by JAI Press Inc.
All rights of reproduction in any form reserved.
ISBN: 0-89232-426-0

transformation processes for creating products and services; (2) in features of firms' products and services; and (3) in firms' managerial-control and other administrative activities. Firms' technologies are continually changing in form and content, as a result of forces internal and external to firms (Sahal, 1982). Here we examine one such force—learning by doing—as a source of changes in firms' process and product technologies. Although research and development is often considered to be the main vehicle used by firms to induce and direct technological change in their products and processes, because learning by doing and research and development (or learning by studying) activities may often be complementary or otherwise related, we compare and contrast these two major ways by which firms' technologies become modified.

Both learning by doing and research and development activities are examined in existing studies but not as related (or potentially related) phenomena and often at macro or industry levels, rather than as emphasized here at the level of individual firms. Not addressed here is firms' capturing new technology (new, that is, to them) by imitating others (adopting existing technology). Diffusion of technological change is in itself a large and controversial subject (Rogers and Shoemaker, 1971). Thus exploring the linkages between diffusion of technological innovation between firms and firms' learning by doing and research and development activities is beyond the scope of this effort.

Economists have been among those who have attempted to incorporate technological change processes into the general fabric of the social sciences. Nelson and Winter (1977) summarized these efforts and pointed to difficulties due to the lack of an integrative theory of technological innovation and change. A number of unfinished tasks remain and Nelson and Winter offer some fruitful approaches to dealing with the unresolved issues. Most notably, they suggest a framework for treating technological innovation as an inherently stochastic process and discuss formulations of theories of technological change that encompass the numerous institutional features and complexities that influence such change.

Technologies imbed knowledge and are reflections of a growing stock of human knowledge. Firms' technologies therefore reflect knowledge acquired from their internal and external sources. An action question for firms' managers is thus: how may they increase their firms' knowledge stocks so as to achieve advantageous technological change? Firms add to their stocks of knowledge—thus accomplishing internal technological changes—by manipulating three internal factors:

1. learning by doing (LbD),
2. research and development (or learning by studying) (R&D), and
3. imitating others (diffusion between firms).

Utilizing these factors, firms modify their market positions and other aspects of their competitive existence. Identifying these three factors illuminates another consideration in studying and managing firms. Firms are not the sole—or at times even the dominant—source of influence over these three sources of change in firms' knowledge stocks (and consequently their internal technological changes). These same factors are influenced directly by factors beyond firms' control— factors such as macro-economic trend and cyclical forces, growth in basic scientific and technical knowledge, social demographics, competitive industry forces, and technological substitution by adjoining industries. Thus firms' efforts regarding technological change are often subject to major external risk and uncertainty, making individual-firm outcomes problematic even when the system in the aggregate—say all firms in the world-wide semiconductor industry in the 1980s—is advancing smoothly and along seemingly predictable lines. The bounds on firms' rationality also introduce risk and uncertainty into internal decisions for inducing technological change, making these internal efforts problematic, and further compounding the difficulties involved in anticipating outcomes. Numerous examples may be given showing obvious linkages between technological discoveries and subsequent product and process achievements, such as the Polaroid Land camera or the Pilkington float-glass process. But all such examples benefit from after-the-fact knowledge of events. Before the fact, outcomes are surrounded by uncertainties and are less obvious.

Thus firms are constrained in their choices for increasing their stocks of process, product, and managerial technical know-how by their internal characteristics and by external macro and micro economic influences. But firms are not technologically impotent. Within these constraints individual firms have varying degrees of latitude for inducing technological change. Directly addressing this latitude, this study compares how firms' learning by doing (LbD) versus their re-

search and development (R&D) behaviors influence their knowledge
stocks—and therefore their manifest internal technological changes.

II. LEARNING BY DOING (LbD) AND RESEARCH AND DEVELOPMENT (R&D)

The connections between industries' and firms' R&D efforts and technological change are the subject of considerable theoretical as well as
empirical study and debate (Nelson, 1981). Economic models generally treat technological advance as the result of an accumulating R&D
capital stock and relate R&D investments to economic variables such
as product demand and factor prices. Studies are often inconclusive
about R&D effects and several issues are unresolved (Piekarz,, 1983;
Scherer, 1983). Studies seldom deal with the inherent uncertainty in
R&D efforts, uncertainty that extends to various facets of technological innovation and development.

As to learning by doing (LbD), studies may note that LbD can be
complementary to or a substitute for research and development efforts
associated with technological change. But technological change studies seldom specifically address LbD or progress due to production
experience. And LbD studies focusing on gains in internal efficiency
generally do not explicitly address issues of technological change
(Dutton and Thomas, 1982). Sahal (1981) and Rosenberg (1982) attempt to deal with learning issues in technological change and provide
illustrative case studies. And their studies at industry or technology
levels open many useful questions about firms' choices.

"Research and development" generally includes numerous activities. "Research" can be seen as the search for scientific and technical knowledge, while "development" is translating this knowledge
into product or process innovations (Abernathy and Rosenbloom,
1969; Gomery, 1983; Link, 1978). The two activities are not always
easily distinguishable; hence, the common heading "research and development." "Learning by doing," on the other hand, refers to
knowledge gained via production experience, whether in the firm itself
or via experience gained by suppliers or, sometimes, users. Both R&D
and LbD involve varied learning processes at different levels of an
organization (Tsuji, 1982). Thus, learning can be viewed as a multidimensional phenomenon, and this discussion attempts to shed light on

how the different dimensions associated with R&D and with LbD interact to induce technological advance.

The development of a new technology is an outcome of learning. Although this learning often stems from deliberate studying, as in R&D activities, it may also stem from experience in using a given technology that gradually leads to evolution of a new technology (Gomery, 1983; Sahal, 1981). Often new technology results from both types of learning. Sahal (1981) distinguishes between "technological learning" and "manufacturing (or operating) learning." The former refers to learning processes advancing a particular technology while the latter describes learning in the operation of a given system of plant and equipment (often employing a number of different technologies). Sahal describes technological learning as a meta-level process in relation to manufacturing learning. He discusses interactions and tradeoffs between the two. The learning described by Sahal as leading to technological innovation is learning as a result of production experience (or sometimes user experience). Here we classify learning processes that occur as a function of production experience as learning by doing (LbD). We thus contrast it from research and development or learning by studying (R&D), as when deliberate R&D is undertaken. Sahal's earlier distinctions are useful in highlighting how these latter two forms of learning may interact to influence technological advances in firms.

R&D is often studied for its impact on technological change, and R&D investments are commonly viewed as technical capital contributing a direct input into the production process (Mansfield, 1964, 1968). Still controversial is measuring output from R&D because technology often seems neither directly nor solely influenced by R&D. If technical progress is considered the desired output from R&D, then measures suitable to assessing the impact of R&D on technical progress would have to control for a number of other mediating factors, including LbD. Factors influencing technical progress appearing in the production process are many. And the manner by which an invention or a breakthrough attained by R&D results in progress at the production level is complex and not generally understood. Patents may be considered a reasonable measure of R&D impact but they measure the rate of inventions, not the direct impact of R&D on progress at the production level. As noted by Kamien and Thirwell (1972: 45): "Moreover, since invention is one step removed from the application of new knowledge to production, a relation between R&D and invention tells us nothing about the impact of R&D on the rate of measured technical progress." In fact, the so-called "one step" combining research and ongoing organizational and operating activities is comprised of several intervening steps influencing how invention affects progress at the production level. And LbD is one key factor in these intervening processes.

In order to understand how R&D and LbD interact to induce technological change and progress in a firm, it is useful to highlight some essential differences between the two sets of activities. Research, development and production can be viewed as a spectrum of activities. As noted earlier, research is the search for scientific and technical knowledge and its successful completion results in an invention or additional knowledge about an existing product or process. Development entails converting the invention or knowledge into innovations suitable for use in a production system. Production involves first breaking-in an innovation and then its systematic use, including coping with bottlenecks and other problems. It is often not clear where one activity ends and the next begins. However, if we view R&D and LbD (production experience) as activities at two ends of this spectrum, certain characteristics associated with the two sets of activities can be distinguished.

R&D constitutes learning that takes place remote from the operating system while LbD (both manufacturing and technological learning) takes place while operating and observing a production system. Learning by studying from R&D involves employees with specialized technical and scientific skills, whereas LbD involves a varied set of employees at the operating level, including direct as well as indirect labor. Typically, R&D is characterized by the involvement of few persons whereas LbD involves many. And although coordination costs are relatively low in R&D activities, they are likely to be high for inducing LbD. By coordination we mean planning, managing and control. R&D is usually characterized by small-scale laboratory or pilot activities while LbD entails commitment to full or at least larger-scale activities. The nature of findings from R&D are likely to be more general and perhaps more theoretical than findings from LbD. This contrast is not surprising, since the latter is tied to a specific mode of operation while the former involves greater degrees of experimentation. It is important to recognize however that LbD also involves experimentation but on the production floor instead of in the laboratory. In addition, the nature of uncertainty and risk as well as the relative returns are likely to be different in R&D activities than in production activities. In the latter, there is invariably some experimentation resulting in modifications of old practices in order to induce greater learning. But these changes or experiments are tied to the existing production process and are generally characterized by lower risk and lower returns than R&D activities. R&D activities involve experimentation significantly beyond present product or process features. Because their uncertainty and risk factors are usually considerably higher than LbD, R&D projects are undertaken in the belief that long-run payoffs in the form of technical progress will be substantial.

III. THE LEARNING BY DOING (LbD) CONCEPT AND ITS CAUSAL DYNAMICS

The learning by doing (LbD) concept is an often loosely defined concept with different meanings based on its use and on academic disciplines. In some areas, LbD is viewed as learning to operate a given system of plant and equipment, that is, improvements attained via operational (or manufacturing) learning within a given technological regime. In a strict sense, what is commonly known as the "manufacturing progress function" describes the above type of learning. The fifty-year history of learning-by-doing studies, however, embraces learning due to several causes, including technological change (Dutton, Thomas and Butler, 1984). In general, the literature on LbD illuminates firms' characteristics with regard to its knowledge-accumulating processes. The term "progress function" is commonly used to denote the LbD concept. A progress function asserts that dynamic improvements in the cost performance of industrial firms can be expected from the growing stock of knowledge they acquire from their production experience. The most common formulation of the progress function is the log-linear form:

$$y = ax^{-b}$$

where
- y = input cost for the xth unit
- x = cumulative number of units produced
- a = input cost for the first unit
- b = progress rate

Four causal factors influencing the slope of the progress function, thus determining firms' dynamic improvements rates, are (Dutton and Thomas, 1982):

1. technological change in capital goods,
2. labor learning (the Horndal effect),
3. scale, and
4. local-system characteristics.

These four main classes of causality are briefly discussed below. Whether progress resulting from the above factors always reflects LbD remains an unresolved issue. Unfortunately, studies of LbD in firms often do not control for the above factors; hence, unravelling their relative effects is often problematic. These factors are discussed separately below with the intent, not of labelling what is and is not learning, but instead to comprehend the dynamics of LbD in production systems and to appreciate the confounding influences which affect LbD in firms.

1. Technological Change in Capital Goods

Technological change in both product and process creates a changing production environment and affects decisions of capital replacement and product and/or process design. In Arrow's (1962) model of the effect of serial capital goods, the emphasis is on technological change in process as the primary cause of the progress phenomenon. Arrow stipulated that the learning by doing (LbD) phemonenon results mainly from technological advances embodied in capital goods; that is, investments in a series of capital goods of improving vintage results in learning by doing in firms. In fact, Arrow suggested using cumulative gross investment as a measure for production experience, instead of cumulative production volume. In Arrow's model, learning is completely embodied in capital goods. Such learning occurs largely in the capital goods supplier firm (and, of course, in the external scientific and technical environment which affects the supplier firm most directly). The supplier firm's learning may or may not be aided by feedback from users of the capital goods.

Technological change in capital goods and its contribution to the progress-function effect has been indirectly studied by Baloff (1966), Conway and Schultz (1959), Crawford and Strauss (1947), and Wright (1936) in the manufacture of steel, electronic goods and aircraft. In general, few studies control for or isolate the effects of serially improving capital goods on the progress function.

2. Labor Learning (The Horndal Effect)

The above name was given by Lundberg (1961) to progress brought about by direct and indirect labor learning, for a given set of capital goods. The Horndal iron works in Sweden had no new investments over a period of 15 years. However, productivity in terms of output per man hour rose about 2% per year. This effect can be attributed to adaptation efforts by labor and technical personnel during production and other cost reducing effects of sustained production of a good.

The Horndal effect was originally thought of as resulting mainly from direct-labor learning. Direct-labor learning is often a key factor in the Horndal effect, but it is only one of many causal factors. Direct-labor learning can be defined as improvement in the performance of tasks of fixed design, largely due to the practice-makes-perfect principle. But such learning by itself tends to asymptote. In fact, adaptation efforts by indirect labor play a significant role in the Horndal effect. These efforts may include adaptation and changes in the form of:

1. additional tooling;
2. changes in process design (including factory layout, manufacturing design of work methods, and other process features);
3. changes in product design (including design flexibility);
4. administrative organization (including plans for supervision, worker training and inspection of work);
5. improved scheduling, inventory management and quality control; and
6. managerial control interventions.

The progress-function literature provides evidence of learning (by direct as well as indirect labor) facilitated by one or more of the above factors. The findings are often not conclusive (and sometimes contradictory) regarding the relative effects of the above factors. Disentangling these factors and their contributions is difficult because of the lack of controls in most studies seeking to determine aggregate learning by doing effects (Dutton and Thomas, 1982).

3. Effects of Scale

Economics of scale are reductions in average costs attributable to increases in scale. However, the dynamics of scale changes and their effects are not generally understood. The traditional thinking about the nature, sources and effects of increases in scale is frequently contradicted by empirical evidence. Gold (1981) reviewed the problems associated with the concept of scale economies and argued for different approaches to grasp the dynamics of scale changes. Scale involves factors which are complex and interactive (Gold, 1981: 28). Progress functions are affected by these factors, including the fact that distinctly different dimensions are associated with the scale of an operation, including: rate of output, scheduled volume of output, absolute size, and physical dimensions of "fixed" factors of production (such as capital goods and average direct labor employment).

While scale itself is a complex subject and while progress functions provide a highly aggregative and empirical view of improvement in input-output ratios, progress function studies do offer evidence of the underlying dynamics present in industrial activities, which more often than not show improvement with experience. Total unit-cost reductions due to increases in production rate and scheduled volume of output appear to contribute to progress as cumulative volume increases. This effect may be due to indivisibilities and to the fact that rate increases are likely to facilitate Horndal-type adaptations and thus contribute to economies. While increased scale may facilitate the Horndal effect, it is important to recognize the separate contributions of these effects. For instance, the Horndal effect can be expected in a production process where the rate of output and other dimensions of scale remain constant for a considerable length of time (Hirsch, 1952; Lundberg, 1961). But often scale effects are confounded with those of learning in empirical studies, making estimates of their separate effects problematic. For instance, Alchian (1959) and Hirschleifer (1962) discuss distinctions between factors of cost reduction in the progress path and note two factors that are often confounded: (1) cost reduction due to increased knowledge resulting from production experience; and (2) cost reduction due to varying techniques of production for different expected volumes of output. The former reduction results primarily from learning by doing while the latter type of cost reduction occurs when production is scheduled in advance and techniques of production vary (for different volumes) in order to exploit economies.

A survey of empirical progress function studies shows a number of instances where significant cost reductions are likely to have resulted from changes in production technique and process design for different expected production volumes (Dutton and Thomas, 1982). Anticipation of expected volume of production makes possible different techniques as well as an integrated output program (instead of several unrelated programs) which in turn yields significant economies. Progress functions do not inherently distinguish between such factors and those due to learning.

4. Local-System Characteristics

After general effects of technology, labor learning and scale are accounted for, a large category of variance in progress functions remains to be explained focusing on the particular characteristics of a given industry and firm.

These local-system characteristics may be general and systemic to an industry as well as idiosyncratic to a firm or parts of a firm. They include such elements as relative ratios of reduction, fabrication, machining and assembly processes, mechanization (or capital intensity), length of manufacturing cycle, variety of inputs-processes-and outputs, product markets (industrial or consumer), etc. (Shen, 1981).

At the individual firm level, these local-system characteristics also include direct and indirect labor adaptations to local conditions, or they may include adaptations made to provide a firm with competitive advantages (Starbuck and Dutton, 1973). Even firms operating in highly similar markets with virtually identical capital goods and labor skills can vary widely in their preferred customers, product mixes, and other operating system characteristics (Child, 1972). Progress functions have been shown to exist under widely different conditions (Dutton and Thomas, 1982; Yelle, 1979), but existing studies do not lend themselves to rigorous examination of the relative effects of local characteristics on the progress path. Taken together, the results of progress studies to date argue for great variety—even uniqueness—in progress rates of individual firms, plants and processes (Conway and Schultz, 1959; Hirsch, 1952; Nadler and Smith, 1963).

In general, progress studies do not provide data on the content or context of the production systems examined and thus do not permit systematic analysis regarding technology, tasks, organization structure and process, and environment. Some theory of organizations exists which does relate and otherwise analyze these factors (Hobbs and Heany, 1977; Lesieur and Puckett, 1969; Miles, 1980; Nystrom and

Starbuck, 1981). Further studies might well reveal associations between how an organization performs with respect to cost improvement from experience and its internal and external characteristics.

The progress function concept is an aggregate one and its underlying causes are not yet well understood. Much of the empirical evidence is difficult to interpret. Improvements arising from a variety of external and internal sources are often presumed to arise from learning. Hence, causal factors and their relative importance are often unclear. But, despite these limitations, the progress-function literature provides useful insights into the learning-by-doing concept.

IV. TECHNOLOGICAL CHANGE PROCESSES IN FIRMS

The following discussion attempts to relate technological change and learning by doing (LbD) in firms based on the available theoretical and empirical evidence for firms' learning by doing. Examination of the influence of the causal factors of progress functions in varied industry and firm settings provides insights into several issues. These issues include the role of these factors in accumulating knowledge capital, their relations to R&D activities and their role in inducing technological change. The dynamics of progress due to one or more of these factors reveals the existence of different types of learning processes within an industry or a technological system. The evidence also reveals the nature of institutional characteristics and policy variables that a firm can influence in order to capture advantages from different types of learning (Dutton and Thomas, 1984). Conditions under which LbD versus learning from deliberate R&D efforts is most effective in knowledge growth and inducement of technological change are also explored. In the process, the limitations of different learning processes and possible tradeoffs between them are studied. The literature provides several case studies highlighting the dynamics of learning in different firms and technological systems.

1.   Technological Change in Capital Goods

Technological change in capital goods has long been recognized as an important source of progress in production processes, and its role in the learning by doing (LbD) phenomenon has been studied by Arrow

(1962), Baloff (1966) and Wright (1936). But what is the nature of the learning that results in improved capital goods? Is it learning by studying (as in deliberate R&D) or LbD (learning from production experience)? At the level of the supplier of the capital goods it is likely to be both of the above. Supplier learning could take the form of product innovation (or change) or process innovation (or change). In the case of product innovation the user of the improved product (capital good) is the customer firm and improvements are reflected in a customer firm's processes or products or both. In the case of process improvements at the capital goods supplier's level, reduced input costs per unit of output for the customer firm may or may not result.

Consider a firm Y—a producer of capital goods used by firm X. If technological progress in firm X is viewed as being affected by technological change in capital goods (supplied by Y) which is partly produced by the Horndal effect (labor learning) within firm Y, then, LbD may play an important role in inducing such technological change in firm Y's product—that is, firm X's capital good. Firm Y's R&D and product innovation could also be influenced by its own learning from production experience—primarily via the Horndal effect. It could also be influenced by firm X's LbD or firm X's experience with using firm Y's products (capital goods). When firm X breaks in, adapts and uses its capital goods in production, and when it learns to eliminate bottlenecks and other production problems, its learning can be fed back to firm Y (Shen, 1981; Von Hippel, 1976). The learning process associated with the Horndal effect in firm X results in operating, and sometimes technical, progress in firm X and, in addition, it can have a direct and significant impact on technological progress in firm Y. This is particularly so if firm X is one of few users or a particularly large and influential user of firm Y's capital goods. That is, the experience of an influential user can be a significant inducement for technological change in capital goods.

LbD can influence such technological change by changes or innovations achieved in firm Y's production process. Or it can do so via experience gained by the user X and fed back to firm Y. Firm X's production experience can indicate essential constraints to the change and modification of a given technology. It can indicate how soon modifications of a technology are likely to reach a stage of diminishing returns and thus guide the supplier firm Y's R&D towards overcoming underlying obstacles. R&D directed towards such goals often turns out to be a starting point for a new and improved technology.

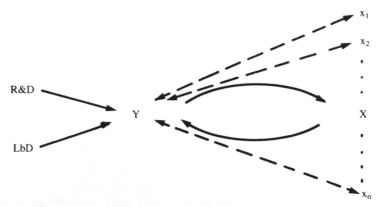

Figure 1.   Learning in Firm Y, and Transfer of Learning between Capital-Goods Firm Y and Customer Firms X and x_i, $i = 1, \ldots n$.

2.   Firms' Internal Learning by Doing (LbD) (as Reflected in the Horndal Effect)

The role of learning by doing (LbD) in technological change at the level of firm X is highlighted by the dynamics of the Horndal effect. (We turn our attention now to the firm in question X). The Horndal effect is one that is mainly endogenous to the firm in nature and origin. A number of factors, listed earlier as causes of the Horndal effect, can be viewed as contributing to improvement in input/output ratios or other performance measures, but not changing the essential technology of the product or process. Factors such as improved scheduling, inventory control, supervision and quality control can usually be classified in a strict sense as improvements in efficiency (and effectiveness) without changing underlying technologies. Other factors such as process redesign, additional tooling and even changes in product design may also contribute to improvement input-output ratios without changing technology but they can also contribute to such improvements via changes in technology. The former type of improvement results when design and tooling changes occur so that different techniques can be employed to produce different volumes of output, to allow substitution of inputs, to allow changing work-force compositions and changing structures of organizational units. But process design changes and product design changes also occur as a result of efforts to eliminate bottlenecks in production and ease the flow of production in order to improve efficiency. Such changes lead to improved techniques which can evolve into minor or major technological changes. In any given system, the causal factors of the Horndal effect can be separated into those which contribute primarily to manufacturing learning and those which contribute to manufacturing as well as technological learning, in the sense defined by Sahal (1981).

Both process and product improvements can occur as a function of the firm's LbD (for a given set of capital goods). Also, due to the loosely defined nature of the LbD concept, process and product improvements have played significant roles in the progress-function effect in many observed instances (Billon, 1966; Conway and Schultz, 1959; Wright, 1936). In other words, what is cause and what is effect is often not clear; and the evidence often reveals these improvements and LbD to be cyclically related. Such improvements or innovations occur in a firm as a result of the research, development and production levels with market forces often playing inducing or mediating roles.

To understand how R&D and LbD interact to induce technological progress, the particular (and unique) characteristics associated with each set of activities must be borne in mind. One essential difference between them is the nature and degree of uncertainty associated with each. Uncertainty is a fundamental feature of many R&D activities. Uncertainty extends: (1) to the expected nature of new developments, (2) to the expense and time involved in new developments, (3) to the future performance of a new technology, and (4) to the issue of which of a set of alternatives will prove most successful (such as placing bets on specific R&D efforts). Modeling the uncertainty inherent in many R&D endeavors has proved difficult in theory. However, it is useful to consider how, if at all, LbD can interact with R&D so that firms can cope with the uncertainties, particularly when making policy.

In their efforts to build a general theory of innovation, Nelson and Winter (1977: 40) note: ''. . . the character of the appropriate institutional structure for the generation, screening and effective exploitation of innovation depends on the underlying technologies, the nature of the demand for the goods and services, and the characteristics of the organizations supplying them.'' The underlying technology of a product or process dictates to a considerable extent what modifications are possible, whether change is brought about by revolutionary or evolutionary means, and what the constraints and limitations to technological change are. Scientists can glean many of the characteristics of a technology by studying and experimenting, as in R&D activities.

It is frequently argued that R&D activities are more effective in the development of certain technologies than others, most notably in those technologies with a strong scientific base. However, LbD can play a significant role both in technologies which do and do not have a strong scientific base.

Understanding technological change as a learning process is aided by Harrison's (1984) distinctions between the processes of science, engineering and technology. Science is the process of investigating natural, physical, and social phenomena, using observation, experimentation and theoretical modeling. Engineering is an applied problem-solving process, focusing on specific, contemporary public and private technological development issues. Technology is the process of producing and delivering goods and services. And technological innovation is the process of discovering how to produce and deliver goods and services more effectively. Harrison (1984) and others have pointed out that technology and engineering drive science as well as vice versa (Gomery, 1983; Greenberg and Goodstein, 1983). And in some fields which are not highly science-based, such as the mechanical engineering trades, production engineering may have a much greater impact than science-based R&D on product design (Amsden, 1983). Thus relating technological change to LbD versus R&D activities is likely to be contingent on the roles that science and engineering play in a particular field of technology, or industry, or in a specific firm's approach (explicit or implicit) to managing technological change.

In technologies which do not require high scientific inputs for their development and operation, it can be argued that LbD is equally and perhaps more effective than R&D, both in learning to use a technology efficiently and in guiding change in the technology. Changes in the underlying technology are guided via learning from production experience (including learning with practice as well as by experimentation on the production floor) and/or learning from user experience in the case of product technology. In technologies with strong scientific bases, separate R&D activity (i.e., experimentation in laboratory conditions separate from the production floor) is often a critical element in technological innovation and change. However, LbD can often play a significant complementary role. And LbD can aid management in dealing with the uncertainty associated with R&D activities.

Experience gained via LbD can serve to target R&D efforts towards technologically feasible alternatives. Learning on the production floor often highlights what features of a product or process cause operating difficulties, where materials or other characteristics can be improved, and equally important, which improvements induced by R&D in a given product or process are likely to yield the greatest progress in the production system. Of course, if R&D is geared towards product innovation, then consumer acceptance and demand potential are crucial factors in the eventual success of the innovation. The learning, adaptation, and experimentation by direct and indirect labor on the production floor that result in the Horndal effect can thus serve as an effective complement to R&D activities in the firm. In order that this be the case, effective links must be forged between the two sets of activities. Then, feedback from the production floor regarding expected performance of a new innovation with respect to rate of operation, ease of operation, scheduling and control problems, capacity problems, and so on, can serve to guide and focus R&D efforts.

Some researchers studying LbD have argued that the benefits of a conventional progress function are obtained primarily from product standardization (Abernathy and Wayne, 1974; Sahal, 1981). We argue that this is not generally the case. Product standardization may aid "manufacturing learning" as defined by Sahal (1981), but the many and varied learning processes which interact to yield the progress function do not indicate that product standardization is necessary in order to attain progress-function benefits. Indeed, the empirical evidence indicates that changes in product design have mixed effects on the progress function. Baloff (1971) and Hirsch (1952) observed production systems (in apparel and machine-tool manufacture respectively) where stability of product design contributed to significant learning. However, other studies reveal instances where regular change in product design and model was a factor that contributed significantly to the observed learning in the system (Conway and Schultz, 1959; Crawford and Strauss, 1947; Middleton, 1945). These above studies were of progress in the assembly of electronic goods and aircraft manufacture. Billon (1966) found, in studies of electronic data processing (EDP) system components and heavy printing machinery, that frequent design changes made predictions of progress difficult, suggesting that the manner in which changes in product design affect progress is unclear.

The difficulty in unravelling which types of learning are aided by product standardizaton and which are aided by changes in product is compounded by the form in which the empirical evidence exists. Typically, studies have confounded many causal factors of the progress observed in production systems. Some of the product-design changes (or process-design changes) that led to high rates of progress no doubt evolved as a result of LbD, particularly in the case of aircraft and apparel manufacture (Baloff, 1971; Middleton, 1945). But often the changes are also likely to have been induced by R&D findings as in electronic goods and aircraft, and in products which involve synthesis such as chemicals or pharmaceutical products. The scarcity of data largely prevents separation of the relative roles of R&D versus LbD in progress of production systems. Instead, we find it more useful to recognize where LbD can aid R&D efforts and focus on areas where links can be forged between the two.

There are other areas where a firm can seek to influence its technological position or gain technological and market advantages via LbD. Firms that develop innovations via R&D or other activities often seek to protect their innovations by patents. Patents increase difficulty for competitors in imitating or copying a firm's innovation, thus enabling firms to increase profits gained from creating innovations. Pavitt (1982: 35), in his study of the relationship between R&D, patenting and innovation, notes that: "Other methods of discouraging imitation involve secrecy, further technological advance based on the firm-specific R&D and skill, influence over suppliers or marketing outlets, and manipulation of standards." When a firm focuses on improving efficiency via LbD, then the resulting benefits (as seen in the Horndal effect) are often in the form of both manufacturing learning as well as technological learning from endogenous sources. Such learning can also be protected as proprietary, as firms such as Michelin and IBM are reputed to emphasize by maintaining high levels of secrecy in plant operations. Therefore, the Horndal effect represents a means for the firm to increase its firm-specific knowledge capital and encourage its R&D activities in directions that are firm-specific and difficult for competitors to pursue or imitate.

Yet another area where LbD can be a useful vehicle for a firm's technological advance arises when a firm is not a leading innovator. Patents may prevent the firm from copying an innovation of a competitor. But there is often a "neighborhood of illumination" (Nelson and Winter, 1977) surrounding an invention or innovation whereby some technological details become publicly known. In such cases, a follower firm can develop a related technology aided by its production experience (LbD) or by its R&D activities. Indeed, firms which choose not to invest heavily in R&D may view the knowledge gained from LbD as significant in aiding and guiding their technological advances. Crown, Cork and Seal provides an example of a firm that for many years seemingly pursued this course of action in the metal container industry with considerable success. In sum, LbD can yield benefits specific to a firm and thus may allow the firm to position itself advantageously with respect to technology in the market place.

The above line of reasoning is not common in the conventional thinking about the progress function or LbD. The progress function describes a dynamic process whereby costs decline as production experience increases. However, it does not stipulate why these cost declines occur or what the relations are between various causes of improvement. Thereby, it is generally believed that progress-function advantages are in the form of increased efficiency (largely in the form of improved input-output ratios) which results from learning to do the same tasks more efficiently with experience. The experimentation, adaptation and the feedback routines in the production system which aid these cost declines are suppressed in many progress-function or LbD studies. These activities also have other benefits besides improving immediate input-output efficiency, such as inducing and aiding firms' technological advances, which are often not acknowledged in the LbD literature.

3. The Interplay of Technological Change in Capital Goods and the Horndal Effect

Examining these two causal factors of progress functions suggests that both may play crucial roles in the technological advances made by firms. They also point to the fact that a firm can benefit from R&D done by other industries as well as from the R&D efforts of other firms in the same industry (as when spillover occurs into the public domain). A firm benefits from the R&D of other industries when the benefits of such R&D are embodied in its capital goods and intermediate product inputs. Progress in a firm stemming from improved capital goods consists largely of benefits from other firms' R&D or LbD, unless the firm is vertically integrated. Economists such as Brown and Conrad (1967), Raines (1976) and Terleckyj (1980) have studied the effects of indirect R&D spending (R&D done by supplier firms) versus the effects of a firm's own R&D spending. Their findings indicate that although both types of R&D are significant in explaining productivity growth of firms in a large number of industries, in general, firms' own R&D spending has less explanatory power than indirect R&D spending. Nelson and Winter (1977) speculate on the reasons for such a finding, noting that firms' own R&D spending is often focused on new product design while process improvements stem from improved capital goods and other inputs (via indirect R&D spending). And process improvements "show up more reliably in increased productivity" (Nelson and Winter, 1977) while problems with price indices make the impact of new product design on productivity harder to assess. Nelson and Winter also claim that since firms' own R&D spending and purchased inputs from R&D-intensive supplier industries are often strong complements, difficulties in correctly modeling the relationship between the two in regression could lead to confusing interpretations.

These findings lead to further speculation on the role of LbD vis à vis technological advance in firms. If technological improvements in capital goods (or other intermediate inputs) have significant productivity implications for a firm (at least as significant as its own R&D spending), then the firm relies heavily on supplier firms for its improvements in process technology. As noted earlier, if a firm is one of a few customers of the supplier firm or is a large and influential customer, then it can influence the technological progress of the supplier firm via user feedback either singly or in conjunction with other customer firms. However, if a firm is one of several customer firms in a highly competitive industry then the firm may have little special influence on its supplier's technological progress. In addition, firms are vulnerable in that technological progress in capital goods is often shared by all customer firms. Hence, a firm cannot rely on this source of progress for advantages in process technology vis à vis its competitors. In such cases, a firm's internal LbD, as reflected in the Horndal effect, is often a powerful means of process improvements in a firm-specific manner, improvements that may give a firm a distinctive niche and related technological as well as market advantages. These benefits of the Horndal effect are mainly due to the fact that the sources of the Horndal effect are internal to firms. And the manner in which firms organize and manage their operations is often a major influence on the degree and kind of benefits gained via the Horndal effect.

4. Scale: A Confounding Factor

Scale is often a major causal factor of progress and is also frequently associated with technological change. Scale has been found to play an important role in the relation between technological change and learning by doing in firms. But it is often confounded with these two phenomena and disentangling its effects often poses a difficult problem.

In a study of technological change and market structure, Mansfield (1983) notes that many economists believe that technological change tends to increase plant sizes and therefore the level of industrial concentration. Other economists such as Blair (1972) have argued that this trend existed prior to World War II, but that subsequently the nature of technological advance has changed so that "centralizing technologies have been displaced and superseded by decentralizing technologies." Mansfield tested Blair's hypothesis in the chemical, petroleum and steel industries and found that in all three industries, scale-increasing innovations far out numbered scale-decreasing innovations. Unfortunately, very few empirical studies have been done on this issue and the existing results should be treated with caution in view of the small number of industries studied. Mansfield's study also indicates that how technological change affects industry concentration is unclear and needs to be studied cautiously. Although the effects on industry concentration are beyond the scope of this discussion, the effects of technological innovation on scale and vice versa are of interest in understanding the processes by which technological change occurs in firms.

Learning by doing (LbD) and adaptation efforts by employees in the course of sustained production often involve solving bottleneck problems and other problems of scheduling, control and flow. Sometimes scale expansions directly result from learning, as often is the case when bottlenecks are eliminated (Lieberman, 1982). And such expansions often imply minor (and sometimes major) technological changes effected within the production system itself. When studying scale in relation to technological change and LbD, the cause and effect relationships are often two-way and difficult to unravel. All three—technological change, LbD and scale—interact and reinforce each other. For instance, increases in output often bring in new capital goods which enable further increases in output; new and technologically improved capital goods may often be scale-increasing as far as the firm's operations are concerned. Scale expansions make possible additional and different kinds of adaptations, thus facilitating LbD via the Horndal effect. And often LbD—whether by direct-labor learning or by process redesign or by additional tooling—results in significant scale changes.

The above four causal factors of progress-function phenomenon illuminate technological-change processes in firms. To further highlight how these factors interact and bring about (or are associated with) technological change in a firm requires recognition and analysis of the firm's strategy with respect to technology, markets and scale and scope of operations.

Two factors may play crucial roles in firms' selection of R&D projects: demand-pull forces and capabilities-push forces. The former are concerned with projects perceived to have good market (demand) potential. These projects are then screened according to cost and feasibility criteria. The latter forces (capabilities-push) are dictated by features of existing technologies or by the visions of scientists or engineers who see certain projects as technologically exciting and feasible. In the case of such projects, management should check for market potential before committing funds to develop new technology (See Pavitt, 1971; Freeman, 1974; Nelson and Winter, 1977 for further discussions of these types of R&D strategies).

When demand-pull forces are the key elements in a given R&D undertaking, then, based on how a firm perceives its future strategic niche it may fund projects leading to scale-increasing innovations. For instance, if a firm believes a large market share can be captured by being a leading innovator where considerable demand potential exists, then, its technological advance is likely to be closely associated with rapid scale increases. That is, its technological changes may be of a scale-expanding nature. And such technological innovations are likely to be followed by construction of new plants or major overhauling of old plants or increases in size and type of capital goods (in what were formerly considered fixed factors of production). The firm can prevent effective imitation by competitors by means of patents or by increasing firm-specific production knowledge via LbD aided by scale advantages.

A capabilities-push R&D strategy means that the firm may have to commit sizeable resources to developing markets for the new technological innovations, particularly if the innovation is of a scale-expanding nature. Here also, the firm may be able to prevent direct copying by patenting its innovation. But often LbD is an effective means of developing firm-specific knowledge thus making it difficult for competitors to pursue the firm effectively. In addition, LbD is often a key instrument in improving production efficiency so that lower costs can be transferred to consumers in the form of lower prices, thus aiding in the development of new markets for the innovation or protecting existing markets.

It would seem in both of the above cases that a firm's R&D and LbD activities need to be reasonably closely linked in order to be effective in the market place. An R&D project involving learning in a laboratory environment distinctly separate from the operating system would no doubt result in technologically exciting product or process innovations. But to transform the innovation into an economic process or product, with sound features and attractive attributes ensuring market demand, requires considerable learning in the operating system via experimentation, adaptation and practice, that is, via LbD (Gomery, 1983). Learning in the operating system in the course of sustained production extends beyond the initial adaptation of new technologies. In fact, such learning is continuous, and its dynamic, ongoing effects can help to steer the direction of additions to a firm's technological knowledge base. The discussion in this and the preceding sections has sought to explore the means by which LbD can steer and otherwise contribute to firms' technological advances and how LbD may interact and complement (or substitute for) firms' R&D activities.

V. TECHNOLOGICAL CHANGE, LEARNING BY DOING AND FIRMS' ORGANIZATIONAL AND MANUFACTURING STRATEGIES

Employing learning by doing (LbD) for bringing about technological changes becomes part of firms' organizational strategies. During Sir Halford Reddish's administration as chairman of Rugby Portland Cement from 1934 to 1971, the company made continuous technical improvement through internal learning an explicit goal (Priedeman, 1969). In other firms, at other times, LbD may be a much more implicit aspect of a firm's managerial policies and organizational culture. Firms' strategies reflect their perceived basic missions (Drucker, 1974). Firms' long-term patterns of strategic decisions reveal these objects and goals, and also their policies and plans for achieving these ends (Christensen et al., 1978). Among these policies and plans are those incorporating firms' implicit and explicit means for achieving

changes in their technologies—whether process, product or organizational. Thus, their technological strategies are revealed in firms' patterns for achieving technological change.

Firms' differing strategies with respect to technological change reflect their differing perceptions of their technological opportunities and
of their strategic situations. Within the major home appliance industry
in the 1960s, competitors pursued a number of different—but nevertheless relatively successful—strategies regarding technological
change (Christensen et al., 1978). General Electric often played the
role of introducing major changes in products' technical features,
while Sears acted as a close follower, quickly adding (to their large-
volume orders to manufacturers) product features proven reliable and
attractive to consumers. Design and Manufacturing Company (D&M),
a non-branded manufacturer supplying a number of other major home
appliance manufacturers as well as national retailers, pursued its own
special technological strategy with respect to product and process developments. After launching a major product innovation by producing
the first successful front loading home dishwasher, D&M maintained
its position as the world's leading supplier of dishwashers, by emphasizing product reliability and lowest-cost production. Another firm in
the major home appliance industry, Tappan Company lacked volume-
cost advantages in manufacturing and distribution; but the firm sought
to otherwise improve its position in the industry by aggressively seeking product innovations, thus attempting to build on its successful
historical performance in home-kitchen gas ranges. Thus, in this period, technology played different roles among major firms in the home
appliance industry, roles that depended heavily on individual firms'
overall product market and manufacturing strategies. Although major
technological forces were present in this industry, these forces did not
impose a common technological imperative upon the firms in the industry, as can be seen from the firms' differing approaches to technological change. In fact, the evidence suggests quite the contrary. A
variety of both feasible and successful technological strategies were
pursued by individual firms; and this was so despite strong competitive
forces in the industry at both manufacturing and distribution levels
(Hammermesh et al., 1978; Hayes and Wheelwright, 1979; Porter,
1980; Skinner, 1969; Starbuck and Dutton, 1973).

Given this background, technological change via LbD constitutes
one implicit or explicit element in firms' strategies. In this context, we
consider how—for specific firms—R&D and LbD may complement

or otherwise contrast in obtaining technological advances and competitive advantages. Analyzing the specific case experiences of two
individual firms further illuminates how firms' technological change
patterns and their basic organizational and manufacturing strategies are
linked.

VI.　INLAND STEEL CORPORATION: TECHNOLOGICAL CHANGES IN CAPITAL GOODS (AND MACHINE CONTROLS)

In 1965, Inland Steel Corporation—the seventh largest steel manufacturer in the United States—introduced a computerized hot strip rolling
mill and thus pioneered the application of computer control to the
rolling of steel (Gabarro and Lorsch, 1979). The impetus for this
innovation was provided by competitive pressures in the form of lower
prices charged by foreign steel manufacturers. The new computerized
80-inch steel mill was expected to be more productive than the older
76-inch mill. Its output averaged 3000 tons per turn by 1967, compared to 1400 tons per turn in the 76-inch mill. The 80-inch mill
employed 680 workers versus 900 employed in the 76-mill and its
operating speed was three times faster than the 76-inch mill. The size
of the 80-inch mill along with its greater needs for maintenance and
downtime caused higher factory overhead, somewhat mitigating its
advantages. Nevertheless, the 80-inch mill was still seen as a technological advance which would enhance productivity considerably
given adequate development and refinement.

The hot strip rolling process was one of a series of processes in the
making of strip steel which was then wound into coils. The strip steel
and coils were manufactured to customer specifications regarding their
dimensions—thickness and width, surface quality and metallurgy. The
hot strip rolling process is outlined in Figure 2.

The most critical function in the hot strip rolling process is the
determination of settings on each finishing mill for every bar processed. Each bar is different and is designated for a specific customer
order. Bad settings result in unacceptable dimensions for the order and
sometimes cause wrecks in the mill, ruining several steel bars and
often damaging the mill itself. Before the advent of the 80-inch mill
and the new computer technology, the determination of settings as well
as prevention of wrecks were mainly the responsibility of the *roller*

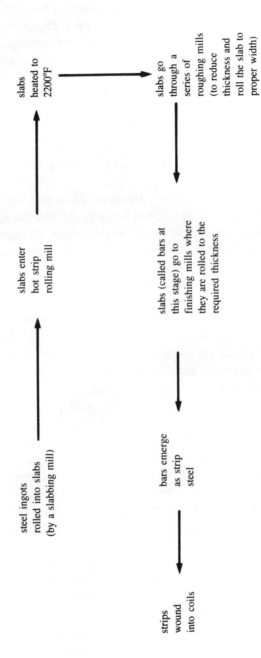

steel ingots
rolled into slabs
(by a slabbing mill)

slabs enter
hot strip
rolling mill

slabs
heated to
2200°F

slabs go
through a
series of
roughing mills
(to reduce
thickness and
roll the slab to
proper width)

slabs (called bars at
this stage) go to
finishing mills where
they are rolled to the
required thickness

bars emerge
as strip
steel

strips
wound
into coils

Figure 2. The Hot Strip Rolling Process.

whose experience and judgement were considered crucial to the effective control of operations. The roller was aided in these tasks by other operators who skillfully controlled line speeds, keeping work flowing and preventing wrecks. With the introduction of the new computerized 80-inch mill, the computer assumed control of settings and production flow. The settings were fed to operators who then made the adjustments while the speed of the lines were also computer controlled.

The nature of the process in the old 76-inch mill together with the custom orders made non-routine work (Perrow, 1967) the norm. The aim of the new technology—the computerized 80-mill—was to make the work routine and automated, requiring fewer workers with lower craft skills. Although the new technology did make some headway in this direction, due to the non-routine work and "uniqueness" of orders, exceptions arose frequently. In fact, the new computer control was often inadequate, creating emergency situations requiring more operators and higher problem-solving skills (including quick diagnoses by electrical and mechanical as well as operating staff). The technological advances in the hot strip rolling process, as well as the progress at the production level, were the result of a series of on-going information exchanges between the R&D staff and the operating crew, contributing to the learning of both. The learning functions that developed and eventually changed the nature of the technology and process are outlined in Table 1.

The sequence of events displayed in Table 1 reveals the two types of learning—learning by studying (R&D) and learning by doing (LbD)—required for breaking in, developing and modifying the computerized hot strip rolling process. It is belaboring the obvious to emphasize that breaking in a new technological innovation and adapting it to the production process requires considerable interaction between R&D and LbD. More subtle and important here is the role of LbD in inducing technological advance in firms, that is, technological progress going beyond merely adapting an innovation into the production process. Inland Steel's experience in this regard illustrates how LbD can induce such technological advance.

The new Inland process-control computer was run off-line at first so as to revise programs, refine settings and add control functions. This adaptation process also involved comparing manual settings and computer control settings, thereby further refining the computer technology. However, LbD in this case went significantly beyond adaptation. LbD contributed to an ongoing effort to improve the new

Table 1.   The 80-Inch Hot Strip Mill with a Computerized Mill-Control System

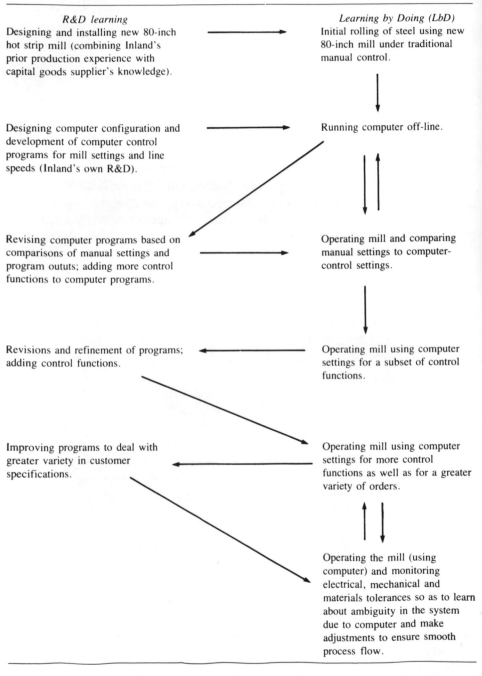

R&D learning

Designing and installing new 80-inch hot strip mill (combining Inland's prior production experience with capital goods supplier's knowledge).

Designing computer configuration and development of computer control programs for mill settings and line speeds (Inland's own R&D).

Revising computer programs based on comparisons of manual settings and program oututs; adding more control functions to computer programs.

Revisions and refinement of programs; adding control functions.

Improving programs to deal with greater variety in customer specifications.

Learning by Doing (LbD)

Initial rolling of steel using new 80-inch mill under traditional manual control.

Running computer off-line.

Operating mill and comparing manual settings to computer-control settings.

Operating mill using computer settings for a subset of control functions.

Operating mill using computer settings for more control functions as well as for a greater variety of orders.

Operating the mill (using computer) and monitoring electrical, mechanical and materials tolerances so as to learn about ambiguity in the system due to computer and make adjustments to ensure smooth process flow.

computerized control programs so as to cope with a variety of order specifications and thus develop a technology more suited to a job-shop type factory rather than one which produced relatively similar outputs. In addition, and equally important, LbD involved monitoring the new computer technology's limitations with respect to electrical and mechanical features of the process. This necessitated greater involvement of electrical and mechanical staff with the operating crew. And Inland gained significant knowledge about the constraints to development of this technology. In studies of learning processes reflected in the evolution of several technologies, Sahal (1981) notes that in many cases technologies eventually become far too complicated to allow further improvements in performance due to endless modifications of the fundamental technology. He adds that the evidence indicates that such stagnation of a technology often proves to be of a temporary nature. Sahal (1981: 10) states: "While obstacles to growth are often real, they seldom last forever. Typically, what would seem to be terminal states of development turn out in reality to be starting points of a great deal of further advances."

The experience of Inland Steel illustrates that LbD allowed the firm to gain considerable knowledge about what modifications were possible in the present technology and what were the constraints to further advance. Whether this form of ongoing learning—together with R&D learning—would eventually result in major technological changes induced by Inland Steel largely depended on the firm's strategy with regard to product and process improvements, and whether its internal institutional characteristics were organized in a manner designed to enhance technological development.

Prior to introduction of the computerized 80-inch hot strip mill, the pace of production flow and the line settings were controlled by the operating crew. The operators were organized by task and horizontally differentiated. The work flow and the roller's functions served as a means of integrating the workers. With the new technology, however, control passed to the computer and away from the workers. The tasks became less craft-oriented and involved interpreting the computer output's guidelines and operating the mills accordingly. Tasks were organized around work stations (i.e., roughing, finishing, etc.) and more direct supervision was introduced. Because of the imperative to learn on the production floor and the resulting feedback from the operating system to the R&D staff, operators often overruled computer guidelines when a job could be done better manually. Although a more-educated, younger work force was brought in to operate the 80-inch mill, they relied heavily on the "mill sense" of the older and more experienced workers in learning mill operations.

As expected, these changes created several organizational problems:

1. developing a closer relationship between the new technical and social systems;
2. alleviating the separation of the operating crew and the R&D staff, particularly since their work was so interdependent (often the contact between the two groups took place only when problems arose); and
3. organizing the information and feedback flows, as well as the operating flow, so that both of the above groups felt responsible, not only for correcting and solving problems as they arose, but for refining and advancing the basic technology.

Prescribing a general solution to organizational problems arising around technological change is beyond the scope of this discussion. But the experience of Inland Steel is illustrative of how characteristics of the local, system organization can significantly affect the direction and scope of technological advance in a firm.

VII. UNITED DIESEL CORPORATION: EMPHASIZING PRODUCT OR PROCESS TECHNOLOGIES

The technological imperatives faced by the United Diesel Corporation (UDC) (Lorsch and Lawrence, 1972) were of a somewhat different nature than the cases discussed earlier. As one of a few large-scale manufacturers of diesel engines for locomotives and other heavy equipment, UDC had enjoyed considerable success in producing customized diesel engines requiring creative product-engineering designs. The large investment required to break into the industry had served as an effective barrier to entry of new firms. However, two developments threatened to change UDC's status in the industry and necessitated changes in the firm's strategy, changes which dictated new technological imperatives. One development was the emergence of new and economic forms of power generation that were competitive with diesel engines. Yet another was the increasingly severe price competition among diesel manufacturers themselves. These developments led UDC to reevaluate its product-market and manufacturing strategies.

UDC's response to these threats mainly took the form of seeking ways to reduce manufacturing costs, delivery lead times and customer maintenance problems. Consequently, the firm's emphasis began gradually to shift away from research, original design and customized production and toward increasing standardization of product designs and greater efficiency in manufacturing. One emergent goal was to reap maximum benefits from manufacturing learning, that is, learning to produce a standardized product more efficiently. But not to be overlooked is that enhancing manufacturing efficiency in this system also involved creative process (and sometimes product) engineering aimed at reducing costs while improving quality. Such creative efforts were largely the responsibility of R&D staff in conjunction with the operating crew, as described below. The R&D staff at UDC included product design engineers as well as specialist design groups (electronic, metallurgical, chemical, etc.) and the draftsmen whose tasks had considerable interdependence with those of the engineers. Table 2 illustrates one of the directions in which the firm seemed to be heading. This approach aims at going down a progress (or learning) curve, via a combination of LbD and R&D efforts focused on designing product and process to lower costs and enhance quality, while keeping product designs relatively stable.

While seemingly headed in the direction indicated in Table 2, man-

Table 2. Strategies at United Diesel Corporation (UDC)

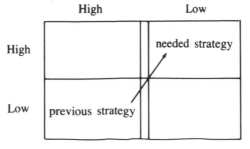

agement frequently seemed indecisive and unsure of how far this commitment should go. The dilemma facing UDC was not atypical in an industry where product (and perhaps process) technologies change fairly rapidly during some periods. While pursuing efficiency and low-cost "learning-curve" advantages, the firm may often be aiming at a moving target. And increasing commitment to standardization may eventually imply reduced technological (as well as other forms of) flexibility. Standardization of product designs could mean a degree of technological stability but could also impede product innovation and might even result in fewer opportunities to achieve greater manufacturing efficiency as standardization reached a stage of diminishing returns. The experience of Ford Motor Company with the Model-T revealed this conflict between standardization versus continuing creative approaches to design (Abernathy and Wayne, 1974). Confronted with this conflict, United Diesel management had to balance its emphasis on increasing manufacturing efficiency with continuing efforts aimed at creative designs of diesel engines, or other power-generation equipment if diversification occurred. Reduced ability for product innovation was likely to be costly to UDC's competitive position, given the growth of several economic forms of power generation. Thus the two-pronged strategy shown in Table 3 may have been indicated for UDC.

Table 3 suggests two approaches: one whereby standardized designs are allowed to remain stable for some length of time (providing stability in product technology) while improving manufacturing efficiency; and another approach emphasizing ongoing experimentation with creative designs that can be produced at relatively low cost. Does this two-pronged approach necessarily imply the separation of R&D and production tasks, whereby the production crew would focus on the former approach while the R&D staff would deal with the latter? Such a separation (R&D versus LbD) would defeat the purpose of a two-pronged strategy. Maintaining and managing the delicate balance between standardization and new creative product designs (for instance, when should minor modifications to an existing standard design be scrapped in favor of a radically new design?) requires task interdependence between the R&D and production crews. The management of UDC faced a struggle in resolving this dilemma, this struggle being reflected in its organizational problems.

The duality of goals discussed above had manifested itself in tensions between the product engineering and the drafting groups. In the

Table 3. UDC's Required Strategies

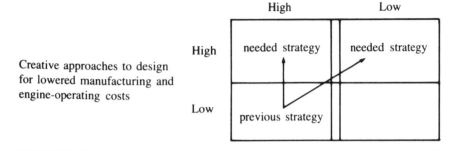

Creative New Diesel Engine Designs

Creative approaches to design for lowered manufacturing and engine-operating costs

past there were close ties between these two groups as they worked together to develop creative designs, with engineers as well as draftsmen proposing new ideas, modifications, and so on. There was low task predictability and problem analyzability in the sense defined by Perrow (1967). The structure of the organization was organic and was designed to facilitate lateral communication, frequent sharing of ideas, and decentralized decision making. Such a structure proved effective due to the goals of creative design engineering and product innovation and it allowed the firm to effectively tap the ideas and creativity of product engineers, draftsmen, as well as operating crews.

Recently, however, the draftsmen and engineers had become unsure of their relative roles and responsibilities. Management was ambivalent about their strategic directions and hence communicated seemingly conflicting goals of efficiency and creative innovativeness to the draftsmen and engineers. The result was considerable tension and competition between the two groups. The product design engineers acted as though designs were their responsibility entirely and expected the draftsmen and operating crews to use routine procedures for carrying out tasks delegated by the engineers. At the same time there was pressure on the draftsmen to continue to display the creativity and innovativeness that had characterized their earlier performance. The resulting confusion within the firm obscured the different strategies that needed simultaneous pursuit for the firm to attain manufacturing efficiency as well as product-market effectiveness via innovative designs and products.

The organizational structure that was necessary had to facilitate creativity on two fronts: (1) in attaining manufacturing process efficiency and high quality with perhaps (though not necessarily) standardized product design and engine operating features; and (2) in inducing product innovation and creating new diesel combustion engineering designs while retaining flexibility for advances of product and process technologies. On both of these fronts, manufacturing efficiency and quality considerations were paramount due to recent severe price competition among manufacturers. To accommodate this two-pronged approach the engineering and drafting group might be split into two groups (each containing both product design engineers and draftsmen). One group—working closely with the operating crews would be involved in experimentation and creative adaptation of the manufacturing process to produce existing designs. The resulting LbD would be a means to attain process efficiencies and low costs while maintaining quality. Such LbD also could indicate the constraints of given diesel combustion engineering designs, as to their future development and ease-of-production characteristics. The second group would serve more specialized R&D functions, such as experimentation with new product designs separate from the production department. Because the manufacturing efficiency of the various designs was a crucial element here, the links and task interdependencies with the former group (particularly with the draftsmen in this case) continued to need careful definition.

VI. CONCLUSIONS

The cases in technological innovation examined here suggest the presence of three important settings for technological change and advance in firms:

1. learning by studying (in a remote scientific or laboratory setting);
2. learning by developing (in a small-scale laboratory or pilot-plant setting); and
3. learning by doing (in a full-scale and complete operating setting).

Learning in firms is a multidimensional phenomenon involving a spectrum of activities, that is, activities ranging from scientific learning (resulting in an invention or added knowledge) to transforming this invention or knowledge into technological progress in production.

Firms have choices regarding what factors they emphasize in order to induce internal learning and thereby technological advances. Of course, firms are constrained in their choices by their internal characteristics and by external influences. In some cases, external constraints may be so severe that they inhibit technological progress no matter what actions firms take to induce such progress. But within these constraints most firms have varying degrees of latitude for internally inducing technological change.

Among the varied learning processes involved in technological advance, R&D and its effects are the most widely studied. But many issues regarding learning in firms remain poorly understood. The path from investing in R&D capital to technological advance involves many types of learning, among which R&D (learning by studying and developing) is only one. Studies of technological change seldom address investments in other forms of learning that firms may have to make in order to induce such change. And these other forms of learning are often a critical influence on rates of technological change and progress in firms.

The bias in academic studies towards the importance of R&D learning in technological progress is transferred to policy matters as well, sometimes in unintended ways. For instance, definitions of firms' activities that constitute research or development for tax purposes used by the Internal Revenue Service are based upon definitions of these activities provided by the National Science Foundation (NSF) (Wasserman, 1982). The studies on which NSF based these definitions, however, may be more cognizant of technological development efforts based on scientific and laboratory work than on learning by doing in operating settings. Such a bias, in academic discussion as well as in government policy, results in inadequate recognition and understanding of the effects of other learning processes that exert major influences on firms' rates of technological advance.

ACKNOWLEDGMENT

This study was supported, in part, by a grant from the Tenneco Fund Program of the Graduate School of Business Administration, New York University.

REFERENCES

Abernathy, W. J. & Rosenbloom, R. S. Parallel strategies in development projects. *Management Science,* 15: 10 (1969), B486–B505.

Abernathy, W. J., & Wayne, K. Limits of the learning curve. *Harvard Business Review*, 52: 5 (1974), 109–119.

Alchian, A. Costs and outputs. In M. Abramovitz et al. (Eds.), *The Allocation of Economic Resources: Essays in Honor of B. F. Haley*. Stanford, CA: Stanford University Press (1959), 23–40.

Amsden, A. H. The division of labor is limited by the rate of growth of the market. Working paper, Harvard University, Graduate School of Business Administration, 1983.

Arrow, K. J. The economic implications of learning by doing. *Review of Economic Studies* 29: (1962), 166–170.

Baloff, N. Startups in machine-intensive production systems. *Journal of Industrial Engineering* XVII: (1966), 25–32.

Baloff, N. Extensions of the learning curve—some empirical results. *Operational Research Quarterly* 22: (1971), 329–340.

Billon, S. A. Industrial learning curves and forecasting. *Management International Review* 6: (1966), 65–96.

Blair, J. *Economic Concentration: Structure, Behavior, and Public Policy*. New York: Harcourt Brace Jovanovich, 1972.

Brown, M., & Conrad, A. The influence of research and education on CES production relations. In M. Brown (Ed.), *The Theory and Empirical Analysis of Production*. Columbia University Press for NBER, 1967.

Child, J. Organizational structures, environment and performance: the role of strategic choice. *Sociology* 6: (1972), 1–22.

Christensen, C. R., Andrews, K. R., & Bower, J. L. *Business Policy*. Homewood, Ill.: Irwin, 1978.

Conway, R. W., & Schultz, A. The manufacturing progress function. *Journal of Industrial Engineering* 10: (1959), 39–54.

Crawford, J. R., & Strauss, E. *Crawford-Strauss Study*. Dayton, OH: Air Material Command, 1947.

Drucker, P. F. *Management*. New York: Harper & Row, 1974.

Dutton, J. M., & Thomas, A. Progress Functions and Production Dynamics. Working paper, New York University, Graduate School of Business Administration, 1982.

Dutton, J. M., & Thomas, A. Treating progress functions as a managerial opportunity. *Academy of Management Review* 9: (1984), 235–247.

Dutton, J. M., Thomas, A., & Butler, J. E. The history of progress functions as a managerial technology. *Business History Review*, 58:(1984), 204–233.

Freeman, C. *The Economics of Industrial Innovation*. Harmondsworth: Penguin, 1974.

Gabarro, J. J., & Lorsch, J. W. Inland Steel Corporation. In J. P. Kotter, L. A. Schlesinger & V. Sathe (Eds.), *Organization*. Homewood, Ill.: Irwin, 1979.

Gold, B. Changing perspectives on size, scale, and returns: an interpretive survey. *Journal of Economic Literature* XIX: (1981), 5–33.

Gomery, R. E. Technology development. *Science* 220:(1983), 575–580.

Greenberg, J. L., & Goodstein, J. R. Theodore von Karman and applied mathematics in America. *Science* 222: (1983), 1300–1304.

Hammermesh, R. G., Anderson, M. J. Jr., & Harris, J. E. Strategies for low market share businesses. *Harvard Business Review* 56: 3 (1978), 95–102.

Harrison, A. J. Science, engineering, and technology. *Science* 223: (1984), 4636 (10 February).

Hayes, R. H., & Wheelwright, S. G. The dynamics of process-product life cycles. *Harvard Business Review* 57: 2 (1979), 127–136.

Hirsch, W. Z. Manufacturing progress functions. *Review of Economics and Statistics* 34: (1952), 143–155.

Hirschleifer, J. The firm's cost function—a successful reconstruction. *The Journal of Business* 35: (1962), 235–255.

Hobbs, J. M., & Heany, D. F. Coupling strategy to operating plans. *Harvard Business Review* 55: 3 (1977), 119–126.

Kamien, M. I., & Thirwell, A. P. Surveys in applied economics. *The Economic Journal* (March, 1972), 11–72.

Lesieur, F. G., & Puckett, E. S. The Scanlon Plan has proved itself. *Harvard Business Review,* 47:5 (1969), 109–118.

Lieberman, M. B. *The Learning Curve, Pricing, and Market Structure in the Chemical Processing Industries.* Doctoral dissertation, Harvard University, Cambridge, MA, 1982.

Link, A. N. Rates of induced technology from investments in research and development. *Southern Economic Journal* 45: 2 (1978), 370–379.

Lorsch, J. W., & Lawrence, P. R. United Diesel Corporation. In Lorsch & Lawrence (Ed.), *Managing Group and Intergroup Relations.* Homewood, Ill.: Dorsey (1972), 243–251.

Lundberg, E. *Produktivitet och Rantabilitet.* Stockholm: P. A. Norstedt and Soner, 1961.

Mansfield, E. Industrial research and development expenditures: determinants, prospects, and relation to size of firm and inventive output. *The Journal of Political Economy* 72: 4 (1964), 319–340.

Mansfield, E. *Industrial Research and Technological Innovation.* New York: W. W. Norton, 1968.

Mansfield, E. Technological change and market structure: an empirical study. *American Economic Review* 73: 2 (1983), 205–209.

Middleton, K. Wartime productivity changes in the airframe industry. *Monthly Labor Review* 61: (1945), 215–225.

Miles, R. H. *Macro Organizational Behavior.* Santa Monica, CA: Goodyear, 1980.

Nadler, G., & Smith, W. D. Manufacturing progress functions for types of processes. *The International Journal of Production Research* 12: (1963), 115–135.

Nelson, R. R. Research on productivity growth and productivity differences: deadends and new departures. *Journal of Economic Literature* XIX: (1981), 1029–1064.

Nelson, R. R., & Winter, S. G. In search of useful theory of innovation. *Research Policy* 6: (1977), 36–76.

Nystrom, P. C., & Starbuck, W. H. (Eds.), *Handbook of Organizational Design,* Volumes 1 and 2, New York: Oxford University Press, 1981.

Pavitt, K. *Conditions of Success in Technological Innovation.* Paris: OECD, 1971.

Pavitt, K. R&D, patenting and innovative activities: a statistical exploration. *Research Policy* 11: (1982), 33–51.

Perrow, C. A framework for the comparative organizational analysis. *American Sociological Review* 32: (1967), 194–208.

Piekarz, R. R&D and productivity growth: policy studies and issues. *American Economic Review* 73:2 (1983), 210–214.

Porter, M. *Competitive Strategy*. New York: The Free Press, 1980.

Priedeman, J. The Rugby Portland Cement Company. In E. P. Learned, C. R. Christensen, K. R. Andrews & W. D. Guth (Eds.), *Business Policy*. Homewood, Ill.: Irwin, 1969.

Raines, F. The impact of applied research and development on productivity. Working paper no. 6814, Washington University, 1976.

Rogers, E. M., & Shoemaker, F. F. *Communication of Innovations*. New York: The Free Press, 1971.

Rosenberg, N. Learning by using. In N. Rosenberg (Ed.), *Inside the Black Box: Technology in Economics*. New York: Cambridge University Press (1982), 120–140.

Sahal, D. Metaprogress Functions. Working paper, New York University, Graduate School of Business Administration, 1981.

Sahal, D. Invention, Innovation and Productivity Growth. Working paper, New York University, Graduate School of Business Administration, 1982.

Scherer, F. M. R&D and declining productivity growth. *American Economic Review* 73:2 (1983), 215–218.

Shen, T. Y. Technology and organizational economics. In P. C. Nystrom & W. H. Starbuck (Eds.). *Handbook of Organizational Design. Volume 1*. New York: Oxford University Press (1981), 268–289.

Skinner, W. Manufacturing—missing link in corporate strategy. *Harvard Business Review* 47: 3 (1969), 136–145.

Starbuck, W. H., & Dutton, J. M. Designing adaptive organizations. *Journal of Business Policy* 3: (1973), 21–28.

Terleckyj, N. E. What do R&D numbers tell us about technological change? *American Economic Review* 70: 2 (1980), 55–61.

Tsuji, K. Scale, technology, and the learning curve. In Buzacott, J. A., Cantley, M. F., Glagolev, V. N., and Tomlinson, R. C. (Eds.), *Scale in Production Systems*. New York: Pergammon Press, 1982.

Von Hippel, E. The dominant role of users in the scientific instruments innovation process. *Research Policy* 5: 3 (1976), 212–239.

Wasserman, B. S. *The Role of Joint Ventures in the Development of Energy Supply Technologies*. Unpublished MBA thesis project, New York University, 1982.

Wright, T. P. Factors affecting the cost of airplanes. *Journal of Aeronautical Sciences* 3: (1936), 122–128.

Yelle, L. E. The learning curve: historical review and comprehensive survey. *Decision Sciences* 10: (1979), 302–328.

Research Annuals and Monographs in Series in
BUSINESS, ECONOMICS AND MANAGEMENT

Advances in the Economics of Energy and Resources
Edited by John R. Moroney, *Department of Economics, Texas A & M University*

Applications of Management Science
Edited by Randall L. Schultz, *School of Management, The University of Texas at Dallas*

Perspectives on Local Public Finance and Public Policy
Edited by John M. Quigley, *Graduate School of Public Policy, University of California, Berkeley*

Public Policy and Government Organizations
Edited by John P. Crecine, *College of Humanities and Social Sciences, Carnegie-Mellon University*

Research in Consumer Behavior
Edited by Jagdish N. Sheth, *School of Business, University of Southern California*

Research in Corporate Social Performance and Policy
Edited by Lee E. Preston, *Center for Business and Public Policy, University of Maryland*

Research in Domestic and International Agribusiness Management
Edited by Ray A. Goldenberg, *Graduate School of Business Administration, Harvard University*

Research in Economic History
Edited by Paul Uselding, *Department of Economics, University of Illinois*

Research in Experimental Economics
Edited by Vernon L. Smith, *Department of Economics, University of Arizona*

Research in Finance
Edited by Haim Levy, *School of Business, The Hebrew University and The Wharton School, University of Pennsylvania*

Research in Governmental and Non-Profit Accounting
Edited by James L. Chan, *Department of Accounting, University of Illinois*

Research in Human Captial and Development
Edited by Ismail Sirageldin, *Departments of Population Dynamics and Policital Economy, The Johns Hopkins University*

Research in International Business and Finance
Edited by H. Peter Grey, *Department of Economics, Rutgers University*

Research in International Business and International Relations
Edited by Anant R. Negandhi, *Department of Business Administration, University of Illinois*

Research in Labor Economics
Edited by Ronald G. Ehrenberg, *School of Industrial and Labor Relations, Cornell University*

Research in Law and Economics
Edited by Richard O. Zerbe, Jr., *School of Public Affairs, University of Washington*

Research in Marketing
Edited by Jagdish N. Sheth, *School of Business, University of Southern California*

Research in Organizational Behavior
Edited by Barry M. Staw, *School of Business Administration, University of California, Berkeley* and L.L. Cummings, *J.L. Kellogg Graduate School of Management, Northwestern University*

Research in Personnel and Human Resources Management
Edited by Kendrith M. Rowland, *Department of Business Administration, University of Illinois* and Gerald R. Ferris, *Department of Management, Texas A & M University*

Research in Philosophy and Technology
Edited by Paul T. Durbin, *Philosophy Department and Center for Science and Culture, University at Delaware.* Review and Bibliography Editor: Carl Mitcham, *New York Polytechnic Institute*

Research in Political Economy
Edited by Paul Zarembka, *Department of Economics, State University of New York at Buffalo*

Research in Population Economics
Edited by T. Paul Schultz, *Department of Economics, Yale University* and Kenneth I. Wolpin, *Department of Economics, Ohio State University*

Research in Public Sector Economics
Edited by P.M. Jackson, *Department of Economics, Leicester University*

Research in Real Estate
Edited by C.F. Sirmans, *Department of Finance, Louisiana State University*

Research in the History of Economic Thought and Methodology
Edited by Warren J. Samuels, *Department of Economics, Michigan State University*

Research in the Sociology of Organizations
Edited by Samuel B. Bacharach, *Department of Organizational Behavior, New York State School of Industrial and Labor Relations, Cornell University*

Research in Transportation Economics
Edited by Theordore E. Keeler, *Department of Economics, University of California, Berkeley*

Research in Urban Economics
Edited by J. Vernon Henderson, *Department of Economics, Brown University*

Research on Technological Innovation, Management and Policy
Edited by Richard S. Rosenbloom, *Graduate School of Business Administration, Harvard University*

Monographs in Series

Contemporary Studies in Applied Behavioral Science
Series Editor: Louis A. Zurcher, *School of Social Work, University of Texas at Austin*

Contemporary Studies in Economics and Financial Analysis
Series Editors: Edward I. Altman and Ingo Walter, *Graduate School of Business Administration, New York University*

Contemporary Studies in Energy Analysis and Policy
Series Editor: Noel D. Uri, *Bureau of Economics; Federal Trade Commission*

Decision Research - A Series of Monographs
Edited by Howard Thomas, *Department of Business Administration, University of Illinois*

Handbook in Behavioral Economics
Edited by Stanley Kaish and Benny Gilad, *Department of Economics, Rutgers University*

Industrial Development and the Social Fabric
Edited by John P. McKay, *Department of History, University of Illinois*

Monographs in Organizational Behavior and Industrial Relations
Edited by Samuel B. Bacharach, *Department of Organizational Behavior, New York State School of Industrial and Labor Relations, Cornell University*

Political Economy and Public Policy
Edited by William Breit, *Department of Economics, Trinity University* and Kenneth G. Elzinga, *Department of Economics, University of Virginia*

Please inquire for detailed brochure on each series

 **JAI PRESS INC., 36 Sherwood Place, P.O. Box 1678
Greenwich, Connecticut 06836**

Telephone: 203-661-7602   Cable Address: JAIPUBL